DYNAMIC MECHANICAL ANALYSIS OF POLYMERIC MATERIAL

MATERIALS SCIENCE MONOGRAPHS

Vol. 1 Dynamic Mechanical Analysis of Polymeric Material (Murayama)

MATERIALS SCIENCE MONOGRAPHS, 1

DYNAMIC MECHANICAL ANALYSIS OF POLYMERIC MATERIAL

TAKAYUKI MURAYAMA

Monsanto Fellow, Monsanto Triangle Park Development Center, Inc.
Research Triangle Park, N.C., U.S.A.

ELSEVIER SCIENTIFIC PUBLISHING COMPANY
Amsterdam – Oxford – New York 1978

ELSEVIER SCIENTIFIC PUBLISHING COMPANY
335 Jan van Galenstraat
P.O. Box 211, Amsterdam, The Netherlands

Distributors for the United States and Canada:

ELSEVIER NORTH-HOLLAND INC.
52, Vanderbilt Avenue
New York, N.Y. 10017

Library of Congress Cataloging in Publication Data

Murayama, Takayuki.
Dynamic mechanical analysis of polymeric material.

(Materials science monographs ; v. 1)
Bibliography: p.
Includes indexes.
1. Polymers and polymerization--Mechanical properties. I. Title. II. Series.
TA455.P58M87 620.1'92 78-2332
ISBN 0-444-41684-6

ISBN 0-444-41684-6 (Vol. 1)
ISBN 0-444-41685-4 (Series)

Printed in The Netherlands

PREFACE

During the past decade, the technology of investigating the dynamic mechanical properties of materials has advanced greatly. The interpretations and applications of dynamic data have stimulated interest from both the practical and the scientific standpoints. Consequently, widespread research work on these topics has established diverse results and principles. There is a need for a monograph which covers the entire field of the dynamic mechanical analysis of polymeric materials.

The purpose of this book is to present the principles of the dynamic mechanical analysis of polymers, the identification and the interpretation of the dynamic mechanical data as properties evaluation, and application of the data to structure-properties relationships and various industrial problems.

Chapter I reviews the principles and related concepts of dynamic mechanical analysis. Chapter II discusses the development of characterization techniques for measuring dynamic mechanical properties. Chapter III is devoted to the interpretations of dynamic mechanical properties at the molecular level; the structural factors include molecular weight, crosslinking, crystallinity, and crystallite morphology. These variables are based on the internal structure of the polymers. However, many polymeric materials are combined as structural units. Examples are the laminated plastics, the fiber-reinforced composites, tires, and fabrics. Dynamic mechanical analyses of these structural units are discussed in Chapter IV.

Chapter V deals with an experimental relationship between dynamic properties and viscoelasticity. External variables which affect dynamic mechanical properties include various deformations (shear, tensile forces, bending, etc.) and environment (liquid or gas). Effects of these variables on dynamic properties are discussed in Chapters VI and VII.

This book is intended as a reference document for chemists, physicists, polymer technologists, and mechanical and chemical engineers engaged in research and development. The practicing dynamicist in industry and government will also find it a useful aid in his work. The book can also be used as a text for senior and graduate-level courses in materials or polymer science and engineering.

The author has, for many years, been an investigator of dynamic mechanical properties of materials at the Monsanto Laboratory in

Research Triangle Park, North Carolina. He thanks his management for encouraging the research project and the publication of this work. He takes this opportunity to thank the numerous colleagues who have assisted him in various ways in preparation of the manuscript, particularly Drs. H. N. Friedlander, A. A. Armstrong, Jr., J. P. Bell, F. C. Brenner, R. Buchdahl, J. H. Dumbleton, V. F. Holland, J. P. Knudsen, E. L. Lawton, D. L. MacLean, T. A. Orofino, B. Silverman, P. J. Stevenson, M. L. Williams, and Prof. C. F. Zorowski, North Carolina State University. The thoughtful comments of Prof. M. Takayanagi, Kyushu University, helped to determine the final content of the manuscript. The author is especially indebted to Dr. J. J. Burke, who painstakingly studied the manuscript and offered numerous suggestions for its improvement.

He is grateful for the permission granted by the following publishers to reproduce material in this book: John Wiley & Sons, New York, Marcel Dekker, New York, and Dr. Dietrich Steinkopff Verlag, Darmstadt, publisher of Kolloid-Zeitschrift & Zeitschrift für Polymere.

He also wishes to acknowledge the help of Mrs. P. M. Tucker and Miss P. L. Haislip in typing and indexing the manuscript.

Research Triangle Park, NC — Takayuki Murayama
December 1977

CHAPTER I

INTRODUCTION TO DYNAMIC MECHANICAL BEHAVIOR

1. INTRODUCTION

Polymeric materials, such as rubbers, plastics, textile fibers, films, foams, and composites, have gained widespread use in industries and in our lives. The production of these high polymers has expanded rapidly during the past decades. As much as any other factor, this widespread use and expansion can be attributed to their versatile mechanical properties. A number of books [1-3] have discussed a wide variety of mechanical behavior of high polymers from both the theoretical and experimental viewpoints. The general principles, as well as useful empirical generalizations of mechanical properties (stress-strain behavior, strength of polymers, creep, and stress-relaxation), have been established in terms of viscoelasticity. The dynamic mechanical behavior has been discussed as a part of these mechanical properties. However, in recent years the technology of investigating the dynamic mechanical properties of materials has advanced greatly. The interpretations and applications of dynamic data have stimulated interest from both the practical and the scientific standpoint. Consequently, widespread research work on these topics has established diverse results and principles.

The commonly used dynamic mechanical instruments measure the deformation of a material in response to vibrational forces. The dynamic modulus, the loss modulus, and a mechanical damping or internal friction are determined from these measurements. The modulus indicates stiffness of material, and it may be a shear, a tensile, or a flexile modulus, depending upon the experimental equipment. The mechanical damping (internal friction) gives the amount of energy dissipated as heat during the deformation.

Polymers are examples of viscoelastic materials, which have some of the characteristics of both viscous liquids and elastic solids. Elastic materials have a capacity to store mechanical energy with no dissipation of energy; on the other hand, a viscous fluid in a nonhydrostatic stress state has a capacity for dissipating energy, but none for storing it. When polymeric materials are deformed, part of the energy is stored as potential energy and part is dissipated as heat. The energy dissipated as heat manifests itself as mechanical damping or internal friction. The internal friction of materials is important, not only as a property index, but also for environmental and industrial applications. Since noise is radiated by the vibration of an object--especially metallic materials, which have small internal friction (0.001 - 0.004)--the application of damping materials to the vibrating surface will convert the energy into heat, which is dissipated within the damping materials rather than being radiated as airborne noise. Amorphous viscoelastic polymers are good damping materials, having high internal friction (0.1 - 0.3). High damping or internal friction is essential in decreasing the effect of undesirable vibration, in reducing the amplitude of resonance vibrations to safe limits, and in all kinds of structures from airplanes to buildings [3-5].

The investigation of the dynamic modulus and internal friction over a wide range of temperatures and frequencies has proven to be very useful in studying the structure of high polymers and the variations of properties in relation to end-use performance. These dynamic parameters have been used to determine the glass transition region, relaxation spectra, degree of crystallinity, molecular orientation, crosslinking, phase separation, structural or morphological changes resulting from processing, and chemical composition in polyblends, graft polymers, and copolymers.

In addition to these structure-property relationships, dynamic mechanical studies have been extended to material composites and structural systems. Considerable attention has been given to the dynamic behavior of structures of composite materials. This is primarily because most aircraft and aerospace structures are subjected to dynamic loadings. Knowledge of the dynamic moduli and properties of composite materials is indispensable to intelligent design with these materials.

2. PRINCIPLE OF DYNAMIC MECHANICAL ANALYSIS

Dynamic mechanical analysis involves the determination of the dynamic mechanical properties of polymers and their assemblies. As a result of this analysis, the relationships between the dynamic properties and the structural parameters (crystallinity, molecular orientation, molecular weight, crosslinking, copolymerization, and plasticization, etc.) and environmental or external variables (temperature, pressure, time, frequency, type of deformation, surrounding atmosphere, humidity, etc.) can be explained.

In order to determine the dynamic mechanical properties (such as the dynamic modulus E', the loss modulus E", and the damping or internal friction, $\tan \delta = E''/E'$), various vibrational methods are used in nondestructive tests. These methods measure the response (deformation) of a material to periodic forces. Thus, the vibrational parameters--amplitude, frequencies, and type of oscillation and wave propagation--become important in this analysis. The physics of vibration and waves has been well developed for these applications [6,7].

Three classes of vibrations are often used in the dynamic mechanical analysis [8].

(1) Free vibrations. If the mechanical properties of the material under investigation are linear (i.e., if its properties are independent of vibrational amplitude), then at a given frequency of oscillation the period and the logarithmic decrement define its mechanical behavior. This is the underlying principle of all free-vibration experiments.

(2) Resonance vibrations. When an oscillating force whose amplitude is fixed, but whose frequency may be varied, is applied to a mechanical system, the amplitude of the resulting vibration passes through a maximum at a frequency which is known as the resonant frequency of the system. The value of the resonant frequency depends on the elastic properties of the system, while the breadth of the resonance gives a measure of the dissipative forces.

(3) Wave propagation [9]. Several types of elastic (or anelastic) waves can propagate in solids. For example, flexural, extensional, and torsional waves can travel along solid rods. In plates, only extensional and flexural waves need be considered. The velocity of propagation of all these waves depends on the elastic constants and the density of the material. Hence, dynamic elastic constants can be calculated from the velocity of the wave propagation in the given medium. When materials display viscous behavior in addition to an elastic response, the parameters describing the dissipative forces can be determined from measurements of the

(a) amount of energy dissipated,

(b) phase relation between stress and strain, and

(c) amplitude ratio of the measured stresses (or strains) at two different locations along the sample.

Experiments of this type have been largely conducted in solid blocks (e.g., concrete, plastics, paper, composites, filaments, etc.) to obtain greater understanding of the dynamic mechanical properties of elastic and viscoelastic solids.

(4) <u>Sinusoidal excitation and response.</u> In viscoelastic studies of polymeric materials the method of sinusoidal excitation and response is very useful [8,10,11]. In this case, the applied force and the resulting deformation both vary sinusoidally with time, the rate usually being specified by the frequency f in cycles/sec or $\omega = 2\pi f$ in radians/sec. For linear viscoelastic behavior, the strain will alternate sinusoidally but will be out of phase with the stress, as seen in Fig. 1.1(B). This phase lag results from the time necessary

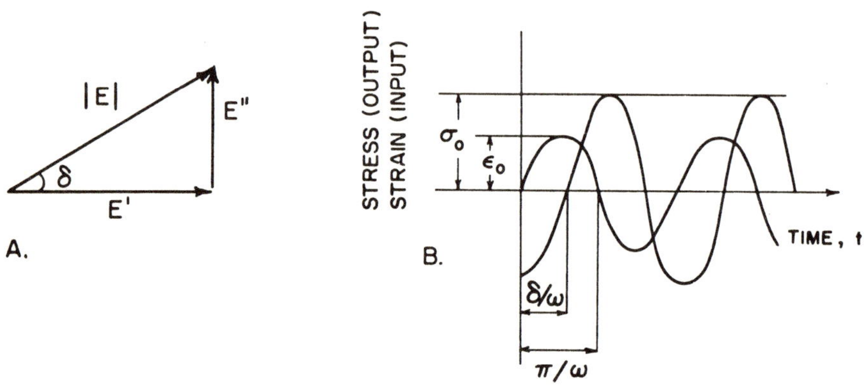

Fig. 1.1. Relations between various parameters used to express the results of a dynamic mechanical measurement.

for molecular rearrangements and is associated with relaxation phenomena [12]. The stress σ and strain ε can be expressed as follows:

$$\sigma = \delta_0 \sin(\omega t + \delta), \qquad (1.1)$$

$$\varepsilon = \varepsilon_0 \sin \omega t, \qquad (1.2)$$

where ω is the angular frequency, and δ is the phase angle. Then

$$\sigma = \sigma_0 \sin \omega t \cos \delta + \sigma_0 \cos \omega t \sin \delta. \qquad (1.3)$$

The stress can be considered to consist of two components, one in phase with the strain ($\sigma_0 \cos \delta$) and the other 90° out of phase ($\sigma_0 \sin \delta$). When these are divided by the strain, we can separate the modulus into an in-phase (real) and out-of-phase (imaginary) component. These relationships are

$$\sigma = \varepsilon_0 E' \sin \omega t + \varepsilon_0 E'' \cos \omega t, \qquad (1.4)$$

$$E' = \frac{\sigma_0}{\varepsilon_0} \cos \delta \quad \text{and} \quad E'' = \frac{\sigma_0}{\varepsilon_0} \sin \delta, \qquad (1.5)$$

where E' is the real part of the modulus, and E" the imaginary part. The complex representation for the modulus as shown in Figure 1.1(A) can be expressed as follows:

$$\varepsilon = \varepsilon_o \exp i\omega t, \tag{1.6}$$

$$\sigma = \sigma_o \exp i(\omega t + \delta). \tag{1.7}$$

Then

$$\frac{\sigma}{\varepsilon} = E^* = \frac{\sigma_o}{\varepsilon_o} e^{i\delta} = \frac{\sigma_o}{\varepsilon_o} (\cos \delta + i \sin \delta)$$

$$= E' + i E'', \tag{1.8}$$

and similarly for the other deformation types. For example, the ratio of the peak stress to peak strain for shear is [8]

$$|G^*| = G'^2 + G''^2, \tag{1.9}$$

where G* is the shear complex modulus, G' is the real part of the modulus and G" the imaginary part, and the phase angle δ is given by

$$\tan \delta = G''/G', \tag{1.10}$$

$$G' = |G^*| \cos \delta \quad \text{and} \quad G'' = |G^*| \sin \delta. \tag{1.11}$$

The real parts of the moduli E' and G', are called the storage moduli, because they are related to the storage of energy as potential energy and its release in the periodic deformation. The imaginary parts of the moduli, E" and G", are called the loss moduli and are associated with the dissipation of energy as heat when the materials are deformed.

These dynamic moduli can also be expressed in terms of a complex compliance

$$J^* = 1/G^* = J' - i J'', \tag{1.12}$$

where J' is the storage compliance and J" is the loss compliance.

The loss tangent tan δ is called internal friction or damping, and is the ratio of energy dissipated per cycle to the maximum potential energy stored during a cycle.

An important feature of dynamic mechanical analysis is the dynamic studies on material composites or structural systems. These materials are not single-component but multicomponent ones, such as fabric-reinforced rubber (for tires), sandwich materials, and laminates. Generally, dynamic analysis for these multicomponent systems is more complex, requiring the concepts of micromechanics and constitutive equations [4,13].

Dynamic mechanical analysis is also related to the fields of acoustics and rheology. Acoustics, viewed broadly, deals with mechanical waves at all frequencies in all substances. Rheology deals with the deformation and flow of matter. Since the dynamic mechanical analysis is concerned with the deformation of materials due to vibrational forces, it relates to both fields.

The applications of acoustics or dynamic analysis are in widespread areas, from seismology (earth waves), oceanography, and architectural engineering to molecular dynamics (crystal lattice vibrations), etc. Table 1.1 shows the frequencies involved in various fields of application. The commonly used frequency range in dynamic mechanical analysis of polymeric materials is from 10^{-2} Hz to 10^{6} Hz.

3. STRESS-STRAIN OSCILLATION

In studies of the response (deformation) of a material to periodic forces, stress, strain, and frequency are considered the key variables. The stress applied to a viscoelastic body results in a dynamic response, linear or nonlinear. For example, if the stress varies sinusoidally with time at a given frequency, then the strain varies cyclically at the same frequency. If the amplitude of the stress is small enough, then the strain is also sinuosidal with time, and the amplitude of the strain is proportional to the amplitude of the stress at a given temperature and frequency: the behavior is linear. However, if the amplitude of the stress is large enough, then the strain varies with the same frequency but is nonsinusoidal; it can have high harmonics, showing nonlinear viscoelastic behavior.

Linear viscoelastic behavior can be explored theoretically using several models [8,10]. The earliest models consisted of a spring and dashpot in series (Maxwell element), a spring and dashpot in parallel (Voigt element), and simple combinations of these basic elements [1].

Each Maxwell or Voigt element is characterized by a relaxation or retardation time defined as $\tau = \eta/E$, which is the viscous damping constant. E is the real modulus. When subjected to a constant stress, the Maxwell element exhibits instantaneous elasticity followed by linear creep. If the loading input is an instantaneous constant strain, the stress-relaxation behavior is an instantaneous elastic response which then decays exponentially. The Voigt element responds to a constant stress input with asymptotically increasing strain. If the stress is removed, the deformation decays as the strain asymptotically approaches zero. Combining these two elements produces a model which exhibits the combined characteristics of the previous

TABLE 1.1

Frequency range of dynamic application

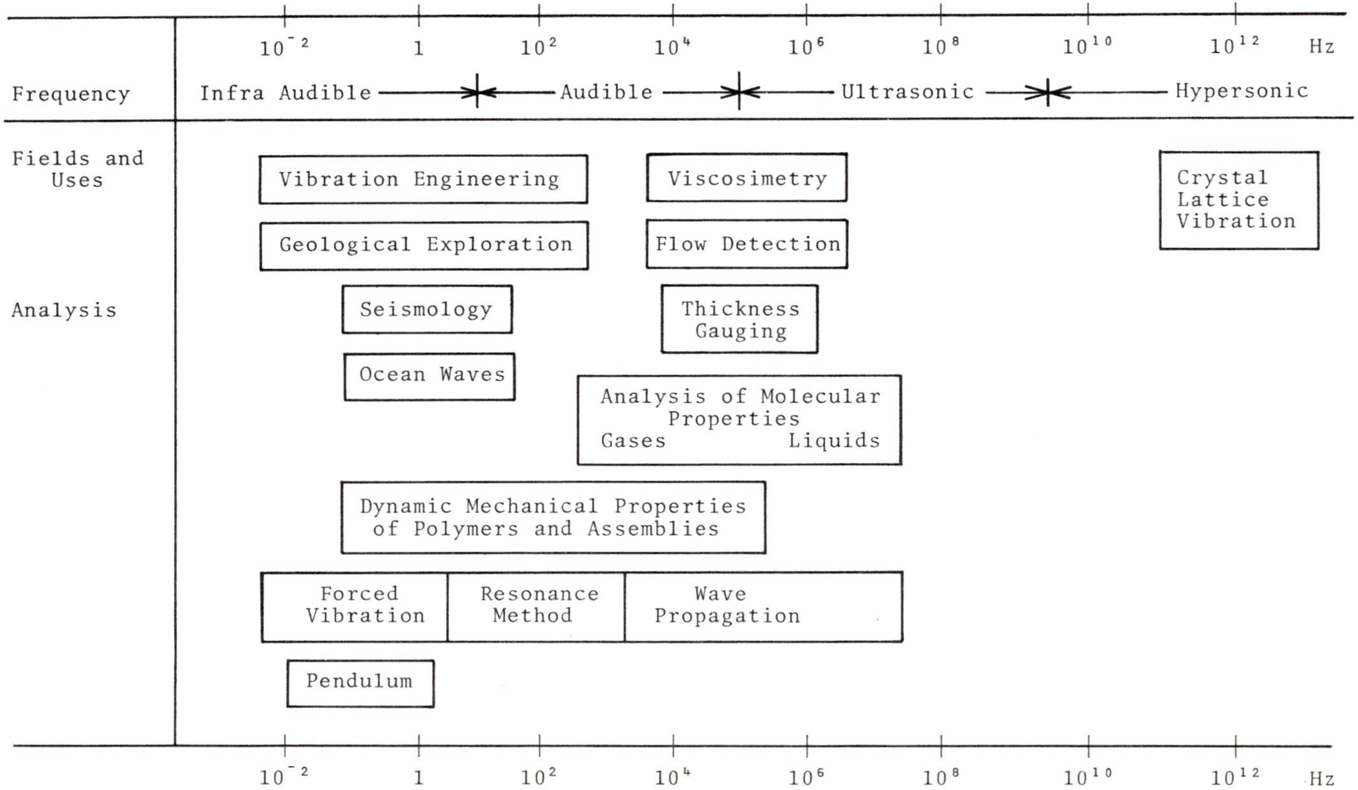

two, i.e., instantaneous elasticity, linear creep, and delayed elasticity. These simple models provide insight into the understanding of such phenomena as creep, relaxation, and memory.

Coupling Maxwell elements in parallel or Voigt elements in series adapts these basic elements to the description of a wider range of phenomena and has led to the concept of a distribution of relaxation times. This assumes that the behavior of a real material can be represented mathematically by such models if they are generalized to contain an infinite number of springs and dashpots in such a manner that the relaxation times τ of the various individual combinations of springs and dashpots are distributed continuously throughout the system. Both of these generalized models have been used for analyzing the mechanical behavior of polymeric materials [8].

Analysis of the behavior of these generalized systems under periodic loading has in turn led to the concept of the complex modulus for characterization of the dynamic mechanical properties of viscoelastic materials. For example, the stress-strain relationship for a generalized Maxwell model containing an infinite number of basic elements subject to a harmonic input of the form $\varepsilon = \varepsilon_o e^{i\omega t}$ with ω constant can be written in the form

$$\sigma = (E' + i\, E'')\, \varepsilon_o e^{i\omega t}, \tag{1.13}$$

where

$$E' = \int_0^\infty \frac{\omega^2\tau^2}{1 + \omega^2\tau^2}\, \phi(\tau)\, d\tau, \tag{1.14}$$

$$E'' = \int_0^\infty \frac{\omega\tau}{1 + \omega^2\tau^2}\, \phi(\tau)\, d\tau, \tag{1.15}$$

and $\phi(\tau)$ is the continuous distribution of relaxation times for the material in question [1].

From equation (1.13) it is possible to define a complex modulus for this model as

$$E^* = \frac{\partial\sigma}{\partial\varepsilon} = E' + i\, E''. \tag{1.16}$$

The individual components of this complex modulus represent the contributions of each Maxwell element in the generalized model summed over the relaxation-time distribution, as indicated in equations (1.14) and (1.15). This complex formulation indicates that the resulting stress is out of phase with the imposed strain.

By rearranging equation (1.13) it is possible to identify an elastic modulus and viscous damping constant in terms of an effective single Voigt element for a viscoelastic material at a fixed frequency, i.e.,

$$\sigma = E'\varepsilon_o e^{i\omega t} + \frac{E''}{\omega}\frac{d}{dt}(\varepsilon_o e^{i\omega t}), \tag{1.17}$$

$$\sigma = E'\varepsilon + \eta\frac{d\varepsilon}{dt}, \tag{1.18}$$

where $\eta = E''/\omega$. Thus the real part of the complex modulus relates directly to the nontime-dependent elastic behavior of the material, while the complex component represents the time-dependent dissipative properties. Both properties are, however, a function of frequency. Figure 1.2 represents the frequency dependence of E' and E" for a Maxwell element [12,16].

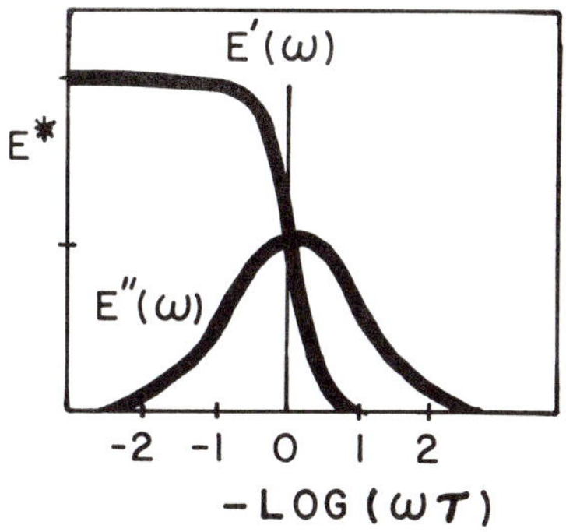

Fig. 1.2. Schematic representation of dispersion E' and absorption E" as a function of angular frequency ω for the single relaxation system.

If the mechanical behavior of viscoelastic material is desired for some loading condition other than constant stress (creep test), constant strain (relaxation test), or harmonic strain variation (complex modulus measurement), the Boltzmann superposition principle is applied. This states that the total strain of a sample is an additive function of the individual strains resulting from each of a sequence of stress applications. Each stress application in the sequence can be handled in terms of one of the simple loading conditions and the particular material properties associated with it.

The stress or strain amplitude and the frequency are related in the heat generation in the polymeric materials during oscillation [3]:

$$Q = \pi E'' \varepsilon_o^2 \qquad (1.19)$$

where Q is the heat dissipated per cycle, ε_o is the maximum value of the strain amplitude during a cycle, and E" is the loss modulus.

Thus, the heat generation in a material is proportional to the square of the amplitude, to the frequency, and to the loss modulus. If a polymeric material is subjected to high frequency, high-amplitude oscillation, the temperature of the specimen is increased. For example energy losses in the rubber compounds and in the fabric of tires under dynamic loading result in a drag component manifesting itself in internal heat generation which is readily observed as an increase in the temperature above ambient [17]. Internal heat generation by oscillation is also a cause of structural changes in polymers. Therefore, in the determination of the dynamic mechanical properties, often we use low levels of strain amplitude and low frequencies. This low strain amplitude is in the linear region of the stress-strain curve. When large strain or stress amplitude is applied to the viscoelastic material, it exhibits increased internal heat generation and nonlinear viscoelastic behavior. Analyses of dynamic response in nonlinear viscoelasticity are generally more complex.

In the nonlinear region the polymers are changed permanently in part of their structure, as in crazing or the formation of microscopic cracks or voids. High strain or stress amplitude at high frequencies can also result in the breaking of bonds, fatigue, and rupture.

4. TIME, TEMPERATURE, AND FREQUENCY

Polymeric properties are temperature-dependent. Plastics are hard and rigid at room temperature (22°C), but become softer at high temperatures (150°C). Similarly, rubbers are elastic and soft at room temperature, but become harder at low temperatures (-30°C). This hardness or stiffness is measured as a modulus, a ratio of stress to strain at a certain stage of deformation. The moduli (Young's modulus dynamic modulus, and relaxation modulus) of polymers are temperature-dependent.

Duration times of forces applied to materials are important variables in the determination of mechanical properties--stress-strain curves, stress relaxation, and creep. For example, the stress-relaxation modulus is a function of time in viscoelastic materials. Modulus measurements over short times result in high values, whereas measurements over long times result in low values. A dynamic mechanical analysis at frequency $\omega = 2\pi f$ is qualitatively equivalent to a transient experiment over a time $t = 1/\omega$.

Since time and temperature are extremely important factors in studies on polymeric materials, much research has been devoted to establishing the principles underlying their influence. The most important concepts are the glass transition [2], time-temperature superposition [8], and the WLF equation [10].

(1) Glass transition T_g and melting point T_m. The properties of polymers change with temperature. In particular, it has been found that the coefficients of thermal expansion of amorphous polymers undergo abrupt changes in the region of the glass transition. The temperature at this abrupt change is defined as the glass transition temperature T_g.

Many amorphous polymers are hard, rigid glasses below the glass transition temperature T_g. This is the glassy region, where the modulus is greater than 10^{10} dyn/cm^2.

In this glassy region, thermal energy is insufficient to surmount the potential barriers for translational and rotational motions of segments of the polymer molecules. The chain segments are frozen in fixed positions. Above the glass transition temperature, the amorphous polymer is soft and flexible. The modulus in this rubbery region is about 10^7 or 10^6 dyn/cm^2. Thus the modulus change between the glassy state and the rubbery state is about four orders of magnitude. This remarkable reduction of modulus is due to the initiation of micro-Brownian motion of the molecular chains from the frozen state with increasing temperature. The thermal energy becomes comparable to the potential energy barriers to segment rotation.

In the glass transition region, the polymer has short-range diffusional motion along with micro-Brownian motion. Segments are free to move from one lattice site to another, and the hard polymer becomes soft and rubbery.

Many mechanical property changes occur in the neighborhood of T_g. The stress-relaxation modulus commonly decreases by about three orders of magnitude, and the creep compliance increases by about three orders of magnitude. In dynamic mechanical studies, the dynamic modulus (storage modulus) decreases rapidly, and the loss moduli and tan δ exhibit maxima.

The physical properties also change near T_g. For example, the line width in a nuclear-magnetic-resonance (NMR) experiment undergoes an abrupt narrowing at this temperature [18]. In polar polymers, the dielectric loss tangent and the imaginary part of the complex dielectric constant pass through a maximum in the neighborhood of T_g. The refractive index shows an abrupt change at T_g [19]. The coefficients of thermal expansion, the mechanical damping (internal friction) [8], and the heat capacity [20] change rapidly with temperature in the glass transition region. All of these effects have been used as measures of T_g. They are all related to molecular dynamics. Since so many different methods exist for the determination of T_g, different values are often obtained. Even the same method will yield different results depending on the time scale. If the experiments are performed over short times, the apparent T_g is raised. If the time scale is longer, the apparent T_g is lowered. Thus, in order to compare T_g values, both the method of measurement and the rate of measurement should be considered. Even if measured consistently, T_g is not a true constant but shifts with time.

The glass transition temperatures for selected polymers are shown in Table 1.2. Values of T_g depend on the chemical structure. Flexibility of the molecular chain, steric hindrance, and bulkiness of the side groups attached to the backbone chain affect the value of T_g [3]. Symmetry also affects the glass transition temperature: an increase in symmetry lowers T_g. Examples are polyvinyl chloride (87°C) and polyvinylidene chloride (-17°C); polypropylene (-10°C) and polyisobutylene (-65°C).

Molecular polarity also influences the value of T_g. Nonpolar polypropylene has a glass temperature of -10°C, moderately polar polyvinyl chloride has a glass transition at 87°C, and highly polar polyacrylonitrile has a transition at 103°C. The glass transition temperature increases with increasing molecular weight and cross-linking. A number of factors related to T_g are described by Nielsen [3]. The theories of the glass transition (free-volume theory, and thermodynamic theory) are summarized by Aklonis [15].

Polymers include not only amorphous rubbers or glasses, but also semicrystalline substances such as nylon, isotactic polypropylene, and polyvinylidene chloride. When crystalline polymers are heated to temperatures much higher than the glass transition temperature, they melt from the solid state to a liquid-like one. The crystallites in a polymer melt over a temperature range. Thus, the melting point T_m is taken as the highest temperature at which crystallinity can be detected in the semicrystalline material. The melting point can be determined from volume-temperature (dilatometric) measurement, differential thermal analysis, and differential scanning calorimetry [21].

TABLE 1.2

Glass transition temperature and melting points of polymers

Polymer	Repeat Unit	T_g (°C)[a]	T_m (°C)[a]
Polyethylene	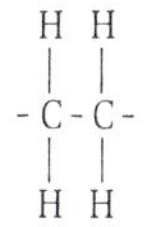	-120 (-130)	137 (141)
Polypropylene (isotactic)	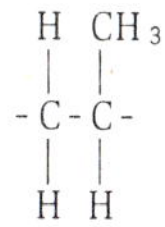	-10	176 (182)
Polyisobutylene	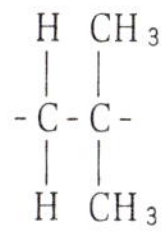	-70 (-73)	128
Poly-1-butene	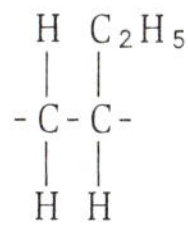	-25	132 (126)
Poly-1-pentene	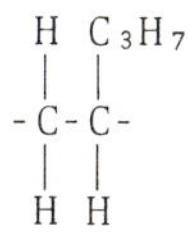	-40	75 (130)
Poly-1-octene	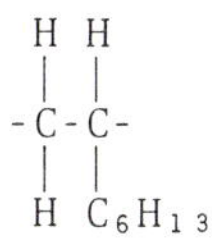	-65	-38
Polyoxymethylene	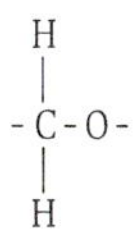	-83 (-50)	181 (198)

TABLE 1.2 (Continued)

Polymer	Repeat Unit	T_g (°C)[a]	T_m (°C)[a]
Polyvinyl methyl ether	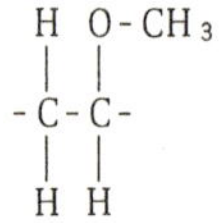	-13 (-20)	144
Polyvinyl ethyl ether	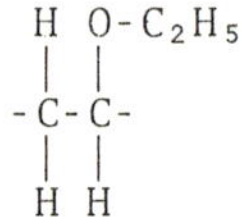	-25 (-42)	86
Polyvinyl-n-butyl ether	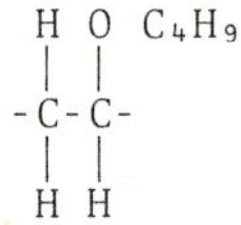	-52 (-55)	64
Polyvinyl isobutyl ether	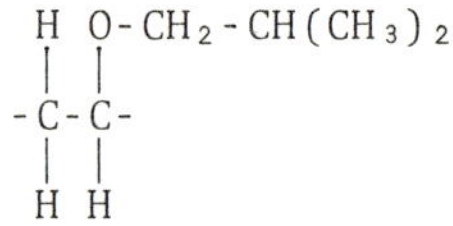	-27 (-18)	115 (165)
Polydimethyl siloxane	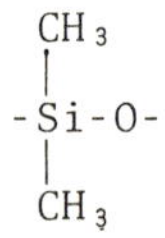	-123	-40
Polystyrene	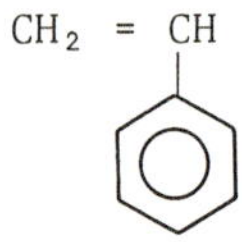	100	240
Polymethyl methacrylate	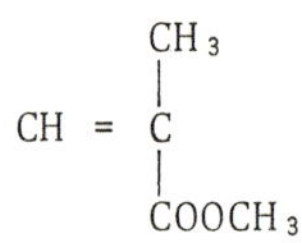	45	160

TABLE 1.2 (Continued)

Polymer	Repeat Unit	T_g (°C)[a]	T_m (°C)[a]
Polyvinyl fluoride	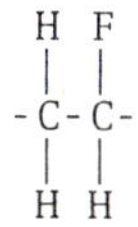	40 (-20)	200
Polyvinyl chloride	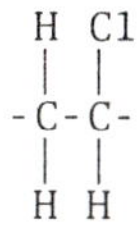	87 (81)	227 (273)
Polytetrafluoroethylene	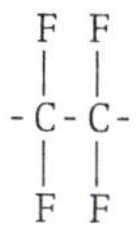	126 (-65)	327 (330)
Polyacrylonitrile (syndiotactic)	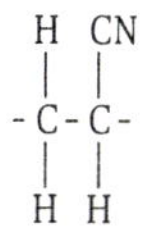	104 (130)	317
Polyvinyl alcohol	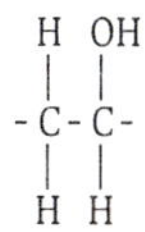	85	870
Polyethylene terephthalate	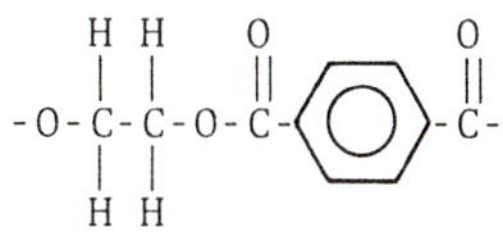	69	265 (325)
Polyhexamethylene adipamide (nylon 66)	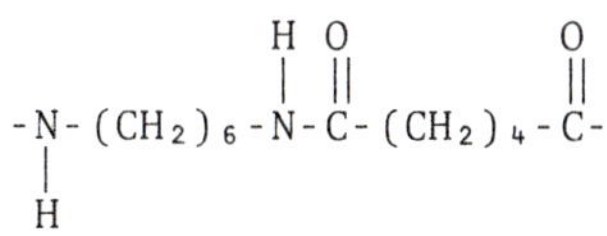	50 (57)	265 (260)

TABLE 1.2 (Continued)

Polymer	Repeat Unit	T_g (°C)[a]	T_m (°C)[a]
Nylon 6 (polycaprolactam)	$-N(H)-(CH_2)_5-C(=O)-$	50 (40)	225 (219)
Poly(bisphenol-A carbonate)	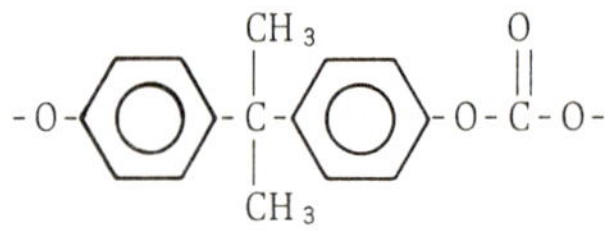	150	267 (220)
Polytetramethylene terephthalate	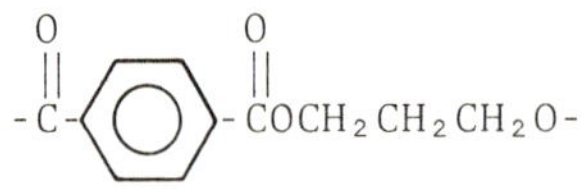	40	265 (325)
Poly-2, 6-dimethyl phenylene oxide	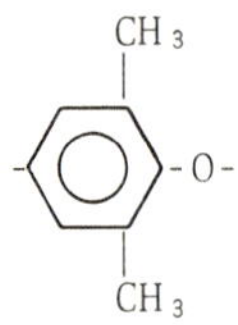	220 (210)	481
Polyhexamethylene sebacamide (nylon 6-10)	$-N(H)-(CH_2)_6-N(H)-C(=O)-(CH_2)_8-C(=O)-$	40 (44)	227 (233)
Polyvinyl tert. butyl ether	$-CH_2-CH(O-C(CH_3)_3)-$ (printed: H, H on first C; H, $O-C(CH_3)_2$ on second C)	88	260

[a]Parentheses indicate alternative values reported in the literature.

With few exceptions, polymer structure affects the glass transition T_g and the crystalline melting point T_m similarly, because cohesive energy and molecular packing affect both the amorphous and the para-crystalline region and thus the temperatures at which the transitions occur. Consequently, T_m and T_g are correlated for many polymers [22,23]. For unsymmetrical polymers the glass transition temperature is about two-thirds of the melting point, when both are expressed in degrees Kelvin. For symmetrical polymers the ratio is roughly one-half.

The melting point of polymers depends upon the molecular weight. The longer the sequence of polymer chain in the crystals, the higher the melting point. The relationship between the melting point and the number average molecular weight $\bar{M}_n$ for polymers with the most probable distribution of molecular weight is shown by Flory [24] to be

$$\frac{1}{T_m} - \frac{1}{T_m^\circ} = \frac{2RM_o}{\Delta H_u \bar{M}_n}. \tag{1.20}$$

Thus the melting point T_m of a material with a given molecular weight is lower than the melting point T_m° of the pure homopolymer of infinite molecular weight. Here the molecular weight of the monomeric unit is M_o, the gas constant is R, and ΔH_u is the heat of fusion per mole of crystalline polymer repeating units. Temperatures are in degrees Kelvin.

The melting point of a polymer may be lowered by solvents or plasticizers and copolymerization. The expression for polymer-solvent systems is [25,26]

$$\frac{1}{T_m} - \frac{1}{T_m^\circ} = \frac{R}{H_u} \frac{V_2}{V_1} (V_1 - X_1 V_1^2), \tag{1.21}$$

where T_m and T_m° are the melting points of the polymer-diluent mixture and pure polymer, respectively, V_1 is the volume fraction of diluent, and V_2 is the molar volume of the polymer repeat unit. X_1 is the polymer-solvent interaction parameter. Thus, the melting-point depression depends on the volume fraction of solvent added and its interaction with the polymer. A similar expression has been developed for copolymers.

The melting points of selected polymers are shown in Table 1.2.

(2) Time-temperature superposition principle. The modulus of a viscoelastic material is a function of time as well as temperature. In fact, there is an equivalence between time (or frequency) and temperature. Leaderman [27] found that in viscoelastic materials data at one temperature can be superimposed upon data taken at a different temperature merely by shifting the creep curves. Stress-relaxation curves (creep curves) of many polymers made at different temperatures are superimposed by horizontal shifts along a logarithmic

time scale to give a master curve. Such curves cover times outside the range easily accessible by experiment.

The relationship between the stress-relaxation modulus E at an arbitrarily selected reference temperature T_o and time t, and the modulus at different temperature T, is expressed as

$$E(T_o,t) = \frac{\rho(T_o)\, T_o}{\rho(T)\, T} E(T,t/a_T), \tag{1.22}$$

where a_T is the shift factor and ρ is the density. The detailed procedures for converting stress-relaxation data at a series of temperatures to a single curve covering many decades of time at some reference temperature have been described by Ferry [8] and Tobolsky [2]. In connection with the shift factor a_T on all amorphous polymers, a well-known relation (the WLF equation) has been developed by Williams, Landel, and Ferry [28]:

$$\log_{10} a_T = \frac{-\,C_1(T - T_g)}{C_2 + T - T_g}. \tag{1.23}$$

The glass transition temperature is chosen as the reference temperature. C_1 and C_2 are constants; $C_1 = 17.4$, $C_2 = 51.6$ for many amorphous polymers. The temperatures are in degrees Kelvin.

a_T is related to the change in viscosity η or the relaxation time τ by $\log_{10} a_T = \log_{10} (\eta/\eta_o) = \log_{10} (\eta/\eta_o)$:

$$a_T = \frac{\eta}{\eta_o} \frac{T_o\, \rho_o}{T\rho}, \tag{1.24}$$

where ρ and ρ_o are the density at temperature T and at the reference temperature T_o, respectively. The glass transition temperature T_g is often chosen as the reference temperature T_o. The WLF equation holds between T_g and 100°C above T_g. It is related to the semi-empirical Doolittle equation for the viscosity η of a liquid [29],

$$\ell n\ \eta = \ell n\ A + B \left[\frac{V - V_f}{V_f}\right], \tag{1.25}$$

in terms of two constants A and B. V is the total volume of the system, while V_f is the free volume available to the system. The viscosity is intimately connected with mobility, which is related to free volume. As the free volume increases, the viscosity rapidly decreases.

The fractional free volume V_f/F is called f; therefore, this equation becomes

$$\ell n\ \eta = \ell n\ A + B\left[\frac{1}{f} - 1\right] . \tag{1.26}$$

It is assumed that above the glass transition temperature, the fractional free volume increases linearly; that is,

$$f = f_g + a_f(T - T_g), \tag{1.27}$$

where f is the fractional free volume at any temperature T above T_g, f_g is the fractional free volume at T_g, and a_f is the thermal coefficient of expansion of the fractional free volume above T_g. In terms of the above equation the Doolittle equation becomes

$$\ell n\ \eta(T) = \ell n\ A + B\left[\frac{1}{f_g + a_f(T - T_g)} - 1\right] \quad \text{for } T > T_g, \tag{1.28}$$

$$\ell n\ \eta(T_g) = \ell n\ A + B\left[\frac{1}{f_g} - 1\right] \quad \text{at } T_g. \tag{1.29}$$

Thus we have

$$\ell n\ \frac{\eta(T)}{\eta(T_g)} = B\left[\frac{1}{f_g + a_f(T - T_g)} - \frac{1}{f_g}\right], \tag{1.30}$$

which simplifies to

$$\log_{10} \frac{\eta(T)}{\eta(T_g)} = \log_{10} a_T = -\frac{B}{2.303\ f_g}\left[\frac{T - T_g}{(f_g/a_f) + T - T_g}\right], \tag{1.31}$$

a form identical to the WLF equation where C_1 is identified with $B/2.303\ f_g$ and C_2 with f_g/a_f. Thus, the form of the WLF equation results from the accurate empirical Doolittle equation and the assumption of a linear expansion of free volume above T_g [15].

(3) Frequency and temperature in dynamic response. In the determination of the dynamic mechanical properties of polymeric materials, the glass transition phenomena and time-temperature superposition principles of viscoelasticity are important factors.

Dynamic methods easily detect glassy-state relaxations and have been extensively applied to their study. Generally, the dynamic measurements are conducted over a frequency range at constant temperature or over a temperature range at constant frequency.

For amorphous or crystalline polymers, a fall in modulus over the glass-transition region is observed, while the loss modulus and the loss tangent tan δ exhibit maxima in this region.

As an example, dynamic data on unoriented amorphous polyethylene terephalate are shown in Fig. 1.3 [30]. The loss tangent tan δ is plotted versus temperature in the range from -70°C to +100°C at a constant frequency of 11 cps. Two relaxation peaks are discernible:

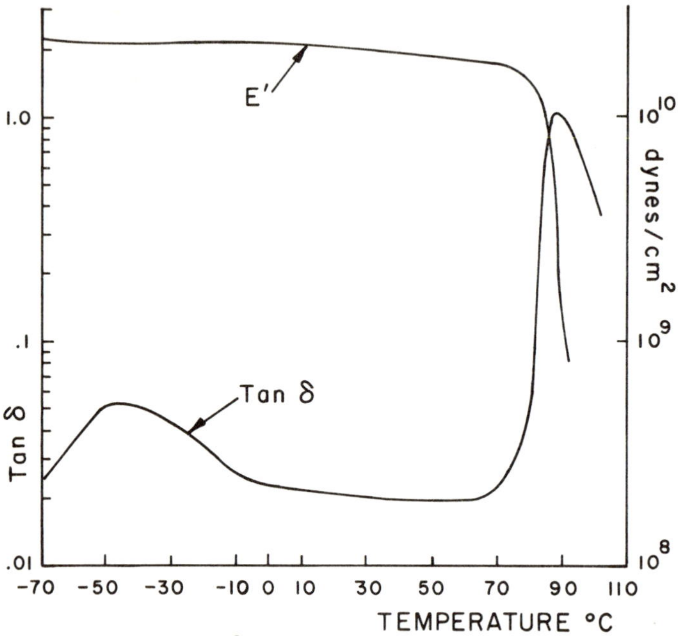

Fig. 1.3. Temperature dependencies of E' and tan δ for unoriented amorphous polyethylene terephthalate.

one corresponding to the glass transition at 85°C, and a secondary peak at -50°C.

The effects of frequency on the dynamic modulus and the loss tangent are illustrated in Fig. 1.4. Since time-temperature equivalence applies to the viscoelastic behavior, an increase in temperature has the same effect on the measured viscoelastic behavior as an increase in time or a decrease in frequency. The dissipation factor (tan δ) goes through a maximum at a frequency which increases with temperature. This behavior can be illustrated by a model with a

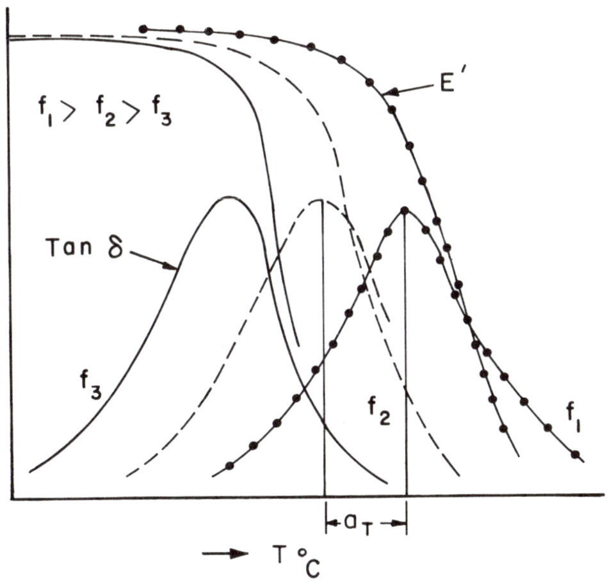

Fig. 1.4. Schematic illustration on effect of frequency f on dynamic modulus E' and loss tangent tan δ.

single relaxation time, where the dissipation factor E"/E' is given by [31,32]

$$\frac{E''}{E'} = 2 \left[\frac{E''}{E'}\right]_{max} \left[\frac{\omega\tau}{1 + \omega^2 \tau^2}\right]. \qquad (1.32)$$

$(E''/E')_{max}$ is the maximum value of the dissipation factor, which occurs at $\omega\tau = 1$. Over a small temperature interval, it may be assumed that the relation between relaxation time and temperature is

$$\tau = \tau_o \, e^{\Delta H/RT}.$$

Combining this equation with the relation $\omega = 1/\tau$, the frequencies at which the damping is a maximum at temperatures T_1 and T_2 (°K) are related by

$$\frac{\tau_2}{\tau_1} = \frac{\omega_2}{\omega_1} = \exp\frac{\Delta H}{R}\left[\frac{1}{T_1} - \frac{1}{T_2}\right], \tag{1.33}$$

where ΔH is the energy of activation. For a decade change in frequency,

$$\frac{\omega_2}{\omega_1} = 10 = e^{2.30} = \exp\frac{\Delta H}{R}\left[\frac{1}{T_1} - \frac{1}{T_2}\right], \tag{1.34}$$

$$\left[\frac{1}{T_1} - \frac{1}{T_2}\right] = \frac{2.303R}{\Delta H}, \tag{1.35}$$

where R is the gas constant. This equation gives the shift in temperature of a maximum damping for a decade change in frequency as a function of the energy of activation. If T_1 is 400°K and ΔH is 1×10^5 calories per mole, the temperature shift $(T_1 - T_2)$ for a factor-of-ten increase in frequency is 7.5°K.

For most polymers, T_g and the maximum in the damping peak are increased about 7°C for a tenfold increase in frequency [3]. Fig. 1.4 is a schematic plot of the modulus and the loss factor tan δ of a typical amorphous polymer as a function of temperature at three frequencies. The shift of the tan δ peak to higher temperatures with increasing frequency occurs in accordance with the time-temperature superposition.

5. STRUCTURE OF POLYMERIC MATERIALS

The dynamic mechanical properties of polymers are strongly influenced by their internal structure. Many kinds of polymeric materials exist: natural polymers (including proteins, carbohydrates, wood, and hair) and synthetic polymers (rubber, films, plastics, and fibers). These polymers are characterized by long chain molecules, high molecular weights, and a variety of chemical compositions and chain conformations.

Studies of the relationships between the properties and the structure of high polymers have been of interest for many years. Since this book is primarily concerned with the dynamic mechanical analysis of polymers, a few elementary concepts of polymer structure will suffice for our purpose. Detailed structural analyses of polymers can be found in the texts by Wunderlich [33] and Geil [34], and in other references [24,35-37].

(1) Chemical composition and polymerization. A polymer is a large molecule built up by the repetition of small, simple chemical units. In some cases, the repetition is linear, much as a chain is built up from its links. In other cases, the chains are branched or

interconnected to form three-dimensional networks. One of the simplest high polymers is polyethylene, which consists of molecules made up of long chains of CH_2 groups. This is an addition polymer and is made by polymerizing the monomer ethylene $CH_2 = CH_2$ to form the polymer $[-CH_2-CH_2-]_n$. The repeat unit of poly(vinyl chloride) is $-CH_2CHCl-$; its monomer is vinyl chloride, $CH_2 = CHCl$. The length of the polymer chain is specified by the number of repeat units in the chain. This is called the degree of polymerization (DP). The molecular weight of the polymer is the product of the molecular weight of the repeat unit and the degree of polymerization. Using poly(vinyl chloride) as an example, a polymer of degree of polymerization 1000 has a molecular weight of 63 X 1000 = 63,000. Most high polymers useful for plastics, rubbers, or fibers have molecular weights between 10,000 and 1,000,000 [38]. Polymers of high molecular weight are called macromolecules. The dividing line between small molecules and macromolecules is given by Staudinger (1950) [39]: Macromolecules have a molecular weight of at least 10,000 or consist of more than 1,000 atoms. The class of macromolecules includes not only linear molecules, but also branched, crosslinked, and three-dimensional molecules. Macromolecules represent the latest and the largest class of matter to be characterized on a molecular level [33].

Monomers become polymers through various polymerization processes. These processes were divided by Carothers [40] and Flory [24] into two groups, known as condensation and addition polymerization or, in more precise terminology, step-reaction and chain-reaction polymerization.

Condensation or step-reaction polymerization is the reaction which takes place between two polyfunctional molecules to produce one large polyfunctional molecule, with the possible elimination of a small molecule such as water. One example is the formation of polyethylene terephthalate (polyester, dacron, or terylene) from ethylene glycol and terephthalic acid:

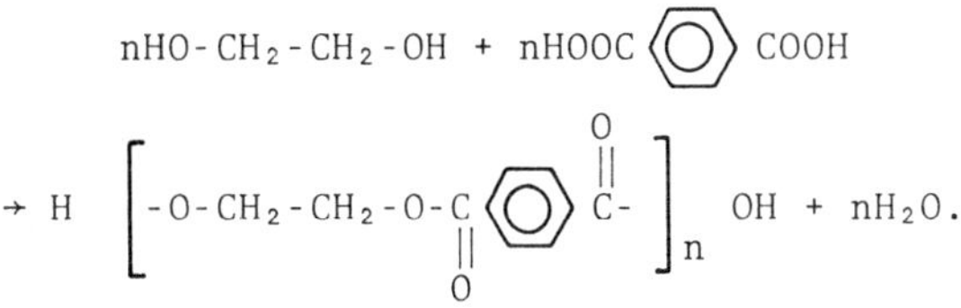

Nylon 66 (a polyamide) can be made from hexamethylene diamine and adipic acid by condensation:

$$nNH_2(CH_2)_6NH_2 + nCOOH(CH_2)_4COOH \rightarrow H\left[NH\ (CH_2)_6\ NHCO(CH_2)_4\ CO\right]_n OH + nH_2O.$$

Addition, or chain polymerization, involves chain reactions in which the chain carrier may be an ion or a reactive substance with one unpaired electron (a free radical). A free radical is usually formed by the decomposition of an unstable material called an initiator

Initiator   Free radical

The free radical is capable of reacting to open the double bond of a vinyl monomer and add to it, with an electron remaining unpaired [38]:

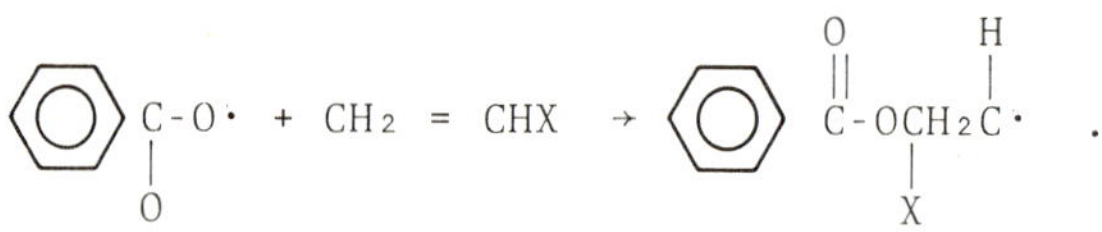

Free radical   Vinyl monomer   Growing polymer chain

In a very short time (usually a few seconds or less) many more monomers add successively to the growing chain. Finally two free radicals react to annihilate each other's growth activity and form one or more polymer molecules:

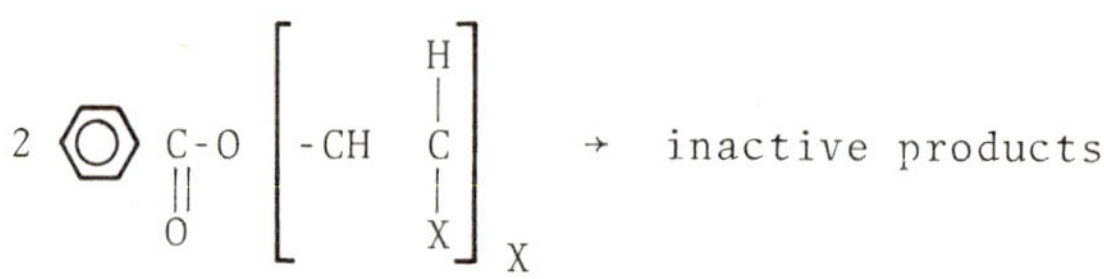

Growing chains   Polymer

(a) Crosslinking and chain-branching. Some polymers have branched chains, as a result of side reactions during polymerization. A secondary chain initiates from a point on the main chain. For example, linear polyethylene consists of long chains of the $-CH_2-$ repeat unit as shown in Fig. 1.5(a), and has high density. Low-density polyethylene possesses usually one long branch per molecule and a larger number of small branches, mainly ethyl $[-CH_2-CH_3]$ or butyl $[-(CH_2)-CH_3]$ side groups, as seen in Fig. 1.5(b). These branched chains lead to considerable difference between the properties of low- and high-density polyethylene.

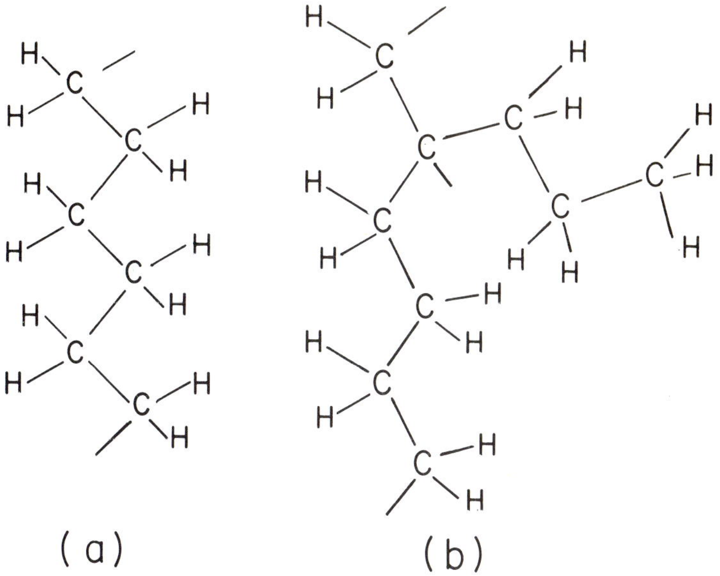

Fig. 1.5. (a) Linear polyethylene; (b) Chain-branched polyethylene.

The crosslinked or network structure is formed when monomers containing more than two reactive groups are polymerized stepwise. This is the situation for rubbers and for the thermosetting resins. The chains are joined by other chains at points along their length to make a structure like the one shown in Fig. 1.6. These reactions can take place during the fabrication of plastics made with thermosetting resins. The crosslinked network extending throughout the final product is stable to heat and cannot be made to flow.

(b) Copolymers. If only one monomer is involved in the synthesis of a macromolecule, it is called a homopolymer. It is also possible to synthesize macromolecules from two or more different monomers, and these are called copolymers. The chemical combination exists in the main chain between two monomers [A] and [B]. A copolymer can be a block copolymer [AAA···] [BB···] or a random copolymer [ABAABAB]. Examples of random copolymers are butadiene styrene and acrylonitrile vinyl chloride.

(c) Molecular weight and molecular-weight distribution. The polymeric material contains molecules having many different chain lengths and molecular weights. The statistical distribution of molecular

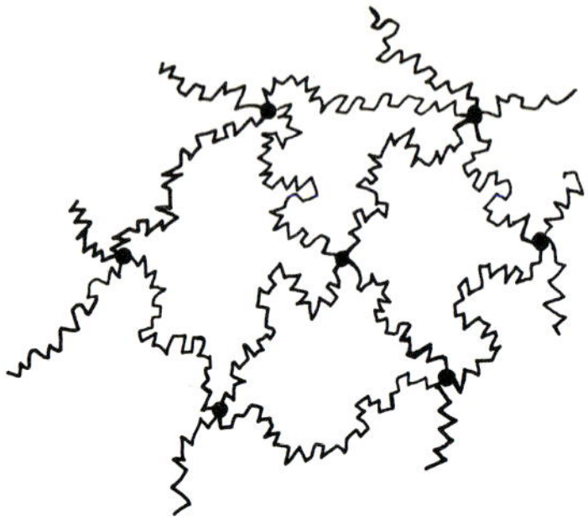

Fig. 1.6. Crosslinked structure.

weights is of importance in determining polymer properties. It can be obtained by gel-permeation chromatography [42].

Since a distribution of molecular weights exists in polymer, the experimental measurement of molecular weight can give only an average value. The number average $\bar{M}_n$ and the weight average $\bar{M}_w$ in dilute solutions of polymers are determined by osmotic-pressure and light-scattering measurements, respectively. For typical polymers the number average lies near the peak of the weight-distribution curve (the most probable molecular weight). The ratio $\bar{M}_w/\bar{M}_n$ is used as a measure of the breadth of the molecular-weight distribution. Values of $\bar{M}_w/\bar{M}_n$ for typical polymers are in the range 1.5 - 2.0 or 20 - 50 [38].

(d) Chemical stereoregularity. In the chemical structure of polymers, two materials which possess the same formula but have different structures are possible. For example, vinyl monomers have the general formula

$$CH_2 = CHR,$$

where R is a monovalent radical such as H, Cl, C_6H_5. In vinyl polymers, successive units may be arranged in the chain in the following sequence:

$$-CH_2-CHR-CH_2-CHR-CH_2-CHR.$$

This is called a head-to-tail arrangement. Alternatively, they may be arranged as follows:

$$-CH_2-CHR-CHR-CH_2-CH_2-CHR.$$

This is called a head-to-head arrangement.

Most vinyl polymers show predominantly head-to-tail substitution. Head-to-head substitution leads to a loss of regularity which can lead to a reduced degree of crystallization, and hence affect mechanical properties [16].

This type of isomerism has been investigated in great detail by Natta and his associates [43]. Three possibilities according to Natta are shown in Fig. 1.7. This figure shows the backbone chain of a vinyl

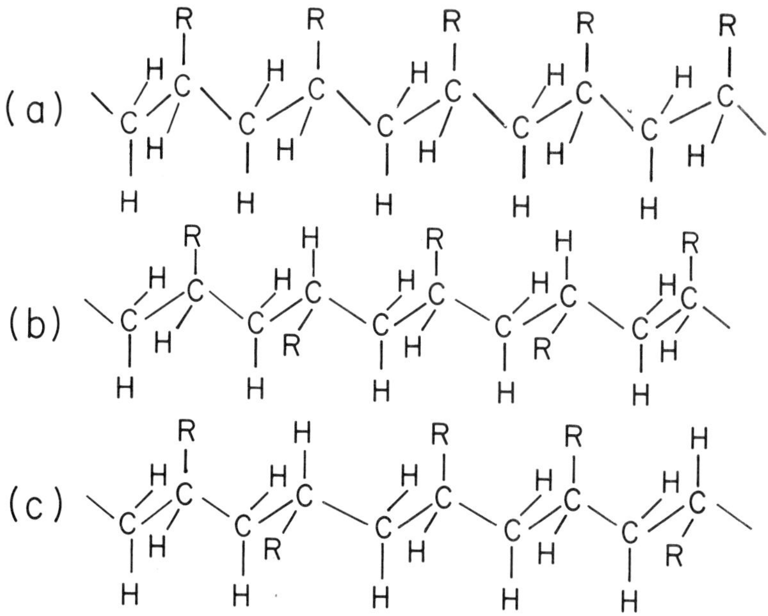

Fig. 1.7. Schematic structure of stereoregulated polymers.

polymer stretched out to the planar zigzag shape. In (a) the substituent radicals R are all above the plan of the backbone; Natta has called this structure isotactic. In (b) the substituents lie alternately above and below the plane of the backbone; this structure is called syndiotactic. Finally, in (c) R occurs randomly above and below the plane of the backbone; this structure is called atactic. The tacticity, or degree of stereoregularity, of a polymer can affect its degree of crystallinity. The isotactic and syndiotactic polymers should be capable of crystallization. The atactic polymer should be amorphous. The degree of stereoregularity is capable of varying over the range from pure isotactic through atactic to pure syndiotactic materials.

(2) Arrangement of the molecules in polymer and morphology. A polymer chain consists of a large number of structural units which can arrange themselves in different ways. In a single macromolecule, monomers can rearrange themselves (intramolecular or micro-Brownian motion). In addition to this motion of single sections or segments of a chain, the entire macromolecule itself moves (intermolecular or macro-Brownian motion). Therefore, the behavior of macromolecular systems is complicated, not only with respect to thermodynamics, but also with respect to chemical reactions.

As described in the previous section, chemical constitution and configuration (including such notions as monomers, chain structure, and chemical stereoregularity) are the primary structural concepts needed for understanding the most important macromolecular compounds. These structural characteristics affect the solubility and the chemical behavior and properties. A polymer chain can exist in a number of different shapes: as a completely extended chain, a random coil, a helix (proteins), or a folded chain. These shapes are types of aggregation states of individual macromolecules (conformations) and are related to the amorphousness or crystallinity of the polymer.

Many polymers, when cooled from the molten state, form a disordered structure, which is termed the amorphous state. This structure is considered to be a random tangle of molecule chains, as seen in Fig. 1.8(a). Rubber, atactic polypropylene (low modulus), polystyrene, polymethylmethacrylate, and melt-quenched polyethylene terephthalate are examples of amorphous polymers. When these polymers are stretched, the molecules are preferentially aligned along the stretch direction. The molecular orientation may be detected by otpical polarization measurements, which give the overall chain orientation.

There are some polymers which crystallize when cooled from the melt or solutions. A crystalline polymer is distinguished from an amorphous polymer by the presence of sharp x-ray lines superimposed on an amorphous halo. Under a polarizing microscope, the presence of polycrystalline aggregates may show as spherulites, as in Fig. 1.9 [44-46].

Spherulites are made up of small crystallites and grow from a nucleus at their center. They consist of narrow chain-folded lamellae (ribbons) growing radially. Since the fibrous crystals (ribbonlike lamellae) are radial, the chains folded within the ribbons are circumferentially oriented [36].

The crystal structures of the crystalline regions can be determined from the wide-angle x-ray diffraction patterns of the polymers. The

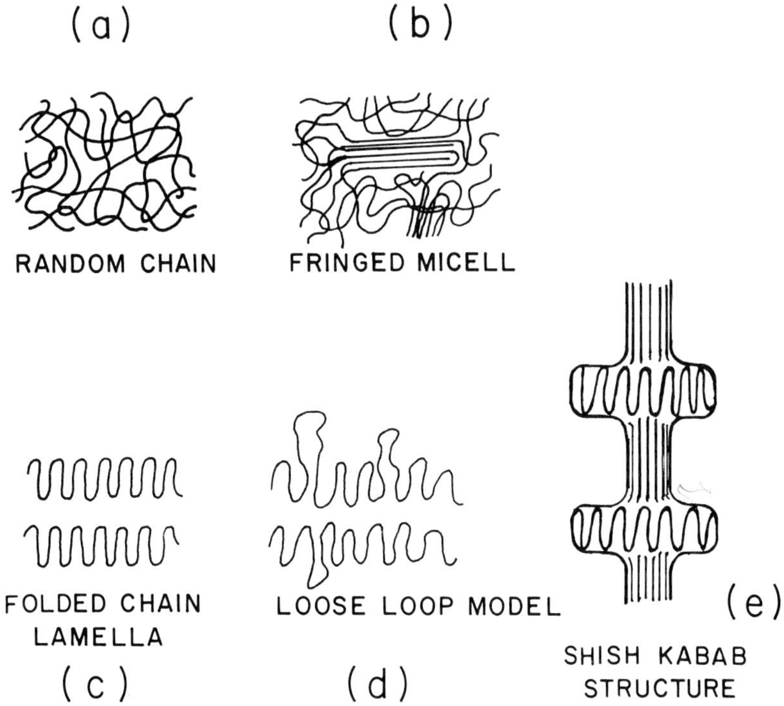

Fig. 1.8. Schematic representation of structure.

sharp x-ray lines lead to the determination of the unit cell, a grouping of atoms which by translation in three dimensions defines the crystal structure [47]. X-ray structural studies on crystalline polymers such as polyethylene, polypropylene, nylon, and polyethylene terephthalate have provided extremely valuable information on their crystalline structure.

In the investigation of semicrystalline polymers, x-ray diffraction patterns show much diffuse scattering due to the amorphous regions. These studies lead to the so-called fringed-micelle model [Fig. 1.8(b)] for the structure of semicrystalline polymers [48]. In this model, the molecular chains alternate between regions of order (the crystallites) and disorder (the amorphous regions), which constitute two phases [49]. The various distortions can be described by the paracrystal model introduced by Hosemann and coworkers [50].

The fringed-micelle model is still in use and may be adequate for the analysis of certain gross properties. However, this model has

Fig. 1.9. Spherulites of polyethylene. (Courtesy of D. C. Felty and V. F. Holland, Monsanto Co.)

undergone considerable revision as a result of the discovery of the chain-folding of molecules in crystals grown from dilute solution [51-53].

Linear polyethylene can form single crystals when carefully grown from very dilute solutions under quiescent conditions. The single crystals are thin plates with lateral dimensions of the order of 10 - 20 microns and of the order of 100 angstrom units in thickness. An electron micrograph of polyethylene single crystals obtained from a dilute solution by Holland is shown in Fig. 1.10. Electron diffraction shows that the molecules are oriented approximately normal to the plane of the lamellae. Since polyethylene molecules are usually of the order of a micron (10,000 Å) in length, Keller [54] postulated that the polymer chains are folded back and forth within the single crystals in layers of uniform thickness in a manner similar to that shown in Fig. 1.8(c), which illustrates perfect crystallographic regularity in folding, with periodic placement of chain folds on the crystal surface. However, chain folds can assume other forms in the crystal, depending on the environment during crystallization as well as on the molecular weight and tacticity of the molecules. The switchboard model and the loose-loop model in Fig. 1.8(d) have been discussed with respect to lack of regularity in the folding [55,56]. An example of interruptions in the regular pattern of folding is interpreted as evidence of edge dislocations in the remarkable moire pattern (Fig. 1.11) from two superimposed single crystals of polyethylene.

These crystals were obtained under quiescent conditions. However, the crystallization of polyethylene from dilute solutions under stress by rapid stirring shows the structure in Fig. 1.8(e), called the <u>shish kebab</u> [57] for its obvious similarity to the popular gastronomic delicacy. A central thread with an extended chain structure is formed under the dynamic stress on the supercooled solution. This nucleus initiates the formation of an overgrowth of ordinary chain-folded lamellae. This discovery prompted the study of crystallization from stressed melts. The results show that lamellae at high stress create an environment favoring numerous short interconnecting tie molecules between lamellae, which account for the elastic properties of certain of these structures [58].

These morphologies (lamellae, spherulite, and tie molecules) are related to the structure of the polymeric materials produced by film blowing [59,60], injection molding [61], and fiber spinning [62-64]. The relationship between these structural parameters and the dynamic mechanical properties will be discussed in Chapter III.

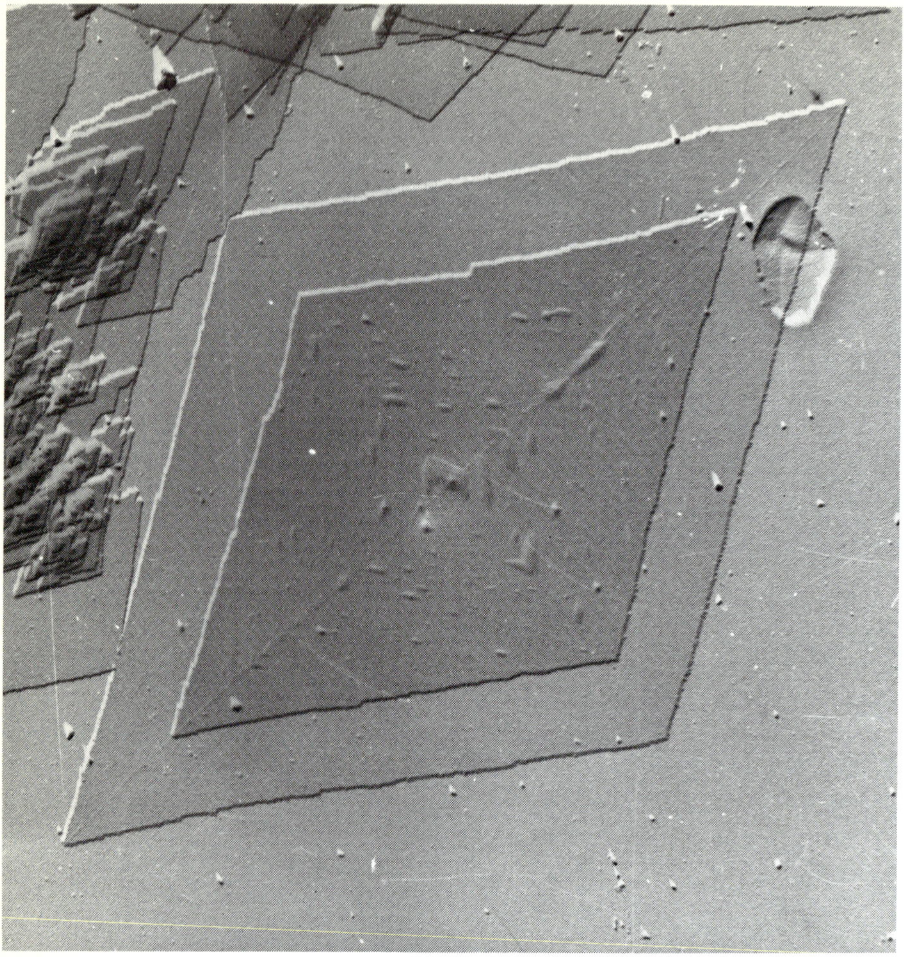

Fig. 1.10. Electron micrograph of polyethylene single crystals, 33,000X. (Courtesy of V. F. Holland, Monsanto Co.)

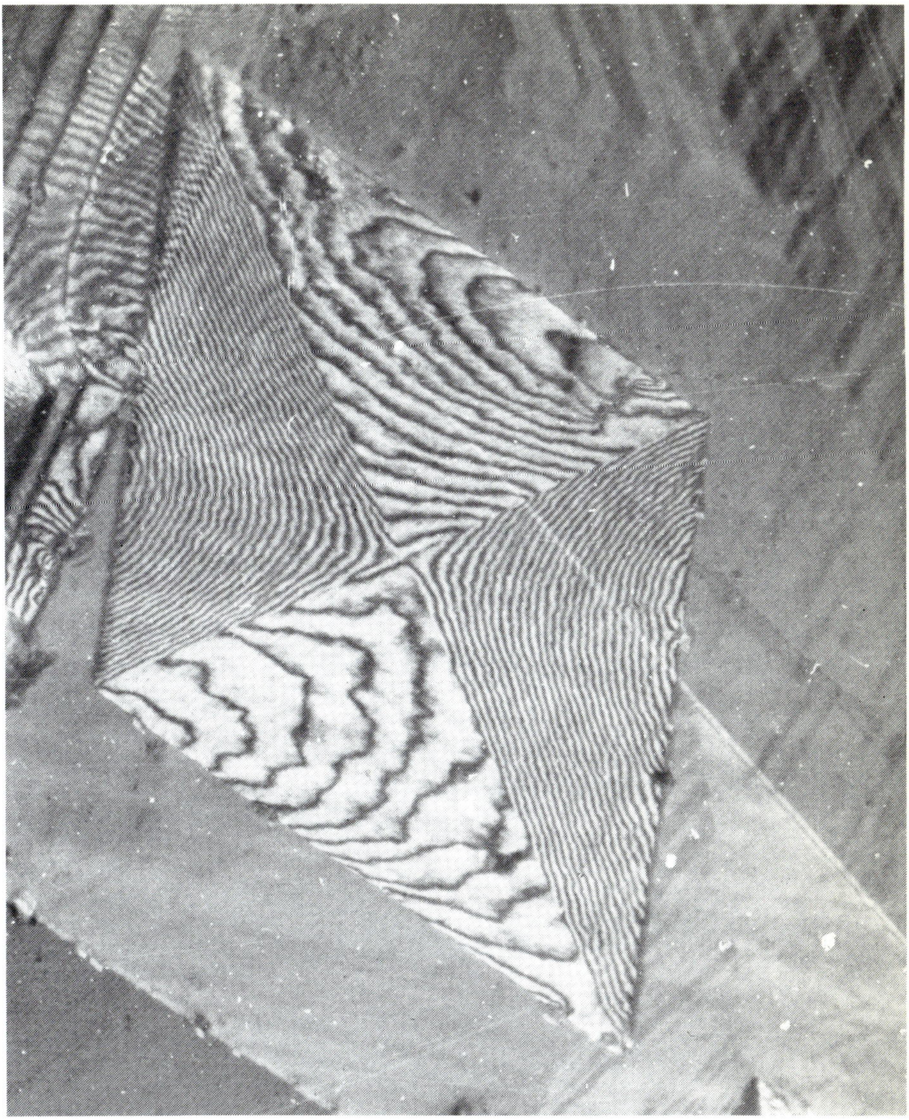

Fig. 1.11. Moiré pattern from two superimposed single crystals of polyethylene. (Courtesy of V. F. Holland, Monsanto Co.)

REFERENCES

1 T. Alfrey, Mechanical Behavior of High Polymers, Interscience, New York, 1948.
2 A. V. Tobolsky, Properties and Structure of Polymers, Wiley, New York, 1960.
3 L. E. Nielsen, Mechanical Properties of Polymers and Composites, Marcel Dekker, New York, 1974.
4 M. P. Blake and W. S. Mitchell, Vibration and Acoustic Measuring Handbook, Spartan, New York, 1972.
5 R. N. Hamme, Harris Handbook of Noise Control: Vibration Damping, McGraw-Hill, New York, 1957.
6 S. Timoshenko, Vibration Problems in Engineering, Van Nostrand, Princeton, 1928.
7 H. J. Pain, The Physics of Vibrations and Waves, Wiley, New York, 1968.
8 J. D. Ferry, Viscoelastic Properties of Polymers, Wiley, New York, 1961.
9 H. Kolsky, Stress Waves in Solids, Clarendon, Oxford, 1953.
10 H. Leaderman, Physics of High Polymers, Tokyo Inst. of Technol., Tokyo, 1957.
11 R. M. Christensen, Theory of Viscoelasticity: An Introduction, Academic, New York, 1971.
12 N. G. McCrum, B. E. Read, and G. Williams, Anelastic and Dielectric Effects in Polymeric Solids, Wiley, London, 1967.
13 S. W. Tsai, J. C. Halpin, and N. J. Pagano, Composite Materials Workshop, Technomic, 1968.
14 A.S.T.M. Special Technical Publication No. 378, Internal Friction, Damping and Cyclic Plasticity, 1964.
15 T. T. Aklonis, W. J. MacKnight, and M. Shen, Introduction to Polymer Viscoelasticity, Wiley, New York, 1972.
16 I. W. Ward, Mechanical Properties of Solid Polymers, Wiley, New York, 1971.
17 S. K. Clark (Ed.), Mechanics of Pneumatic Tires, N.B.S. Monogr. 122, 1971.
18 J. A. Sauer and A. E. Woodward, Rev. Mod. Phys., 32 (1960) 88.
19 R. H. Wiley and G. M. Brauer, J. Polym. Sci., 3 (1948) 647.
20 B. Ke, Newer Methods of Polymer Characterization, Interscience, New York, 1964.
21 R. S. Porter and J. F. Johnson (Eds.), Analytical Calorimetry, Plenum, New York, 1969.
22 R. G. Beaman, J. Polym. Sci., 9 (1942) 470.
23 R. F. Boyer, J. Appl. Phys., 25 (1954) 825.
24 P. J. Flory, Principles of Polymer Chemistry, Chap. 13, Cornell U. P., Ithaca, 1953.
25 P. J. Flory, Trans. Faraday Soc., 51 (1952) 848.
26 L. Mandelkern, Chem. Rev., 56 (1956) 903.
27 H. Leaderman, Elastic and Creep Properties of Filamentous Materials and Other High Polymers, Textile Foundation, Washington, D.C., 1943.
28 M. L. Williams, R. F. Landel, and J. D. Ferry, J. Am. Chem. Soc., 77 (1955) 3701.
29 A. K. Doolittle and D. B. Doolittle, J. Appl. Phys., 31 (1959) 1164.
30 T. H. Dumbleton and T. Murayama, Kolloid-Z. Z.-Polym., 220 (1967) 1; 41-48.
31 N. G. McCrum, J. Polym. Sci., 27 (1958) 555.
32 A. E. Woodward and J. A. Sauer, The Physical Properties of Polymers, Soc. Chem. Ind. Monogr. No. 5, p. 245, Macmillan, New York, 1959.
33 B. Wunderlich, Macromolecular Physics, Vols. 1, 2, Academic, New York, 1973.

34 P. H. Geil, Polymer Single Crystals, Interscience, New York, 1963.
35 M. Takayanagi, Viscoelastic Properties of Crystalline Polymers, Mem. of the Fac. of Eng., Kyushu Univ., Vol. XXIII (1963), No. 1.
36 E. S. Clark, Structure of Crystalline Polymers, Polymeric Materials, Am. Soc. for Metals, Metals Park, Ohio, 1973.
37 R. T. Samuels, Structured Polymer Properties, Wiley, New York, 1974.
38 F. W. Billmeyer, Jr., Textbook of Polymer Science, Interscience, New York, 1965.
39 H. Standinger, Organische Kolloidchemie, 3rd edn., Vieweg, Braunschweig, 1950.
40 H. Mark and G. S. Whitby (Eds.), Collected Papers of Wallace Hume Carothers on High Polymeric Substances, Interscience, New York, 1940.
41 B. Follmert, Polymer Chemistry, Springer, New York, 1973.
42 M. F. Vaughan, Nature, 188 (1960) 55.
43 G. Natta and P. Corradini, J. Polym. Sci., 39 (1959) 29.
44 C. W. Bunn and T. C. Alcock, Trans. Faraday Soc., 41 (1945) 317.
45 W. M. D. Bryant, J. Polym. Sci., 2 (1947) 547.
46 R. G. Vadimsky, H. D. Keith, and F. J. Padden, Jr., J. Polym. Sci., A-2, 7 (1969) 1367.
47 C. W. Bunn, Chemical Crystallography, Clarendon, Oxford, 1946.
48 O. Gernross, K. Hermann, and W. Abitz, Z. Phys. Chem., 10B (1930) 371.
49 H. A. Stuart and Ann Stuart, N. Y. Acad. Sci., 83 (1959) 1.
50 R. Hosemann and R. Bonart, Kolloid-Z., 152 (1957) 53.
51 P. H. Till, J. Polym. Sci., 24 (1957) 301.
52 A. Keller, Philos. Mag., 2 (1957) 1171.
53 E. W. Fischer, Z. Naturforsch., 12a (1957) 753.
54 A. Keller, Makromol. Chem., 34 (1959) 1.
55 P. H. Geil, Chem. and Eng. News, 16 Aug. 1965, pp. 72-84.
56 R. F. Boyer, Paper presented to the U.S.-Japan Joint Seminar on Polymer Solid State, Cleveland, 1972.
57 A. J. Pennings, J.M.A.A. Van der Mark, and A. M. Keil, Kolloid-Z. Z.-Polym., 237 (1970) 336.
58 E. S. Clark, in "Structure and Properties of Polymer Films", (R. W. Lenz and R. S. Stein, eds.), Plenum Press, New York, 1973.
59 P. H. Lindenmeyer and S. Lustig, J. Appl. Polym. Sci., 9 (1965) 227.
60 V. F. Holland, J. Appl. Phys. 35 (1964) 3235.

CHAPTER II

INSTRUMENTATION

1. INTRODUCTION

The development of characterization techniques for measuring dynamic mechanical properties has been an important activity in the area of dynamic mechanical analysis. An accurate measurement of the response of a polymeric material to vibrational forces requires many instrumentational considerations. A very large number of techniques have been developed by using different vibrational principles at various frequencies. The methods are free vibration, resonance forced vibration, nonresonance forced vibration, and the wave- or pulse-propagation technique. The frequency for each method has a limited range. For instance, the frequencies of forced vibration are about 10^{-3} Hz to 10^{2} Hz. In the case of wave propagation, the frequency range used is from 10^{3} Hz to 10^{7} Hz.

Since polymeric materials exhibit a wide variety of shapes and moduli (polymer melts, rubbers, plastics, and fibers) the instrument is often specialized for one type of material. It is very difficult to develop one instrument which has the capability of characterizing all types of polymers over a wide range of frequency and temperature. Many instruments are described in the literature [4-22]. Ferry and Nielsen have summarized these instruments in their books, [1-3].

During the past decade, however, the technology for investigating the dynamic mechanical properties of materials has advanced greatly due to new electronic circuits, improved transducers, signal receiving systems, and computerized controlling systems. A great quantity and variety of research has been carried out by these means [1,23-25]. In this chapter, some of the representative instruments will be discussed.

2. RHEOVIBRON

The Rheovibron is designed to measure the temperature dependence of the complex modulus (E^*), dynamic storage modulus (E'), dynamic loss modulus (E''), and dynamic loss tangent ($\tan \delta$) of viscoelastic materials at specific selected frequencies (0.01 to 1 Hz, 3.5, 11, 35, 110 Hz) of strain input. This instrument was developed by

Takayanagi. It was unique because it was designed to read tan δ directly, greatly simplifying and speeding the characterization of a material. The Rheovibron was commercialized by Toyo Measuring Instruments Co.

During measurement, a sinusoidal tensile strain is imposed on one end of the sample, and a sinusoidal tensile stress is measured at the other end. The phase angle δ between strain and stress in the sample is measured by a direct-reading method [22]. The instrument uses two transducers for detection of the complex dynamic modulus (ratio of maximum stress amplitude to maximum strain amplitude) and the phase angle δ between stress and strain. From these two quantities, the real part (E') and the imaginary part (E") of the complex dynamic tensile modulus (E*) can be calculated.

Figure 2.1 shows the arrangement of transducers and sample. In this case, one end of the sample was driven at a certain frequency

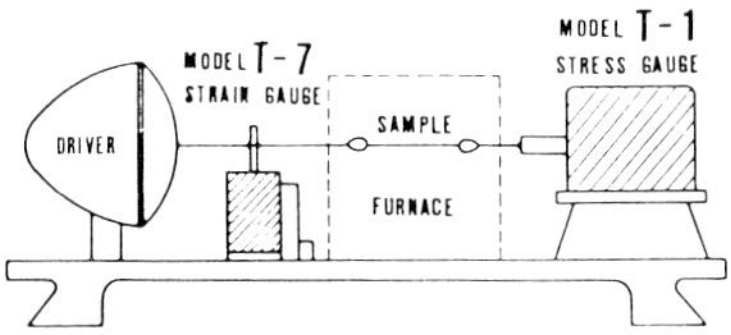

Fig. 2.1. Arrangement of transducers.

by means of a magnetic coil. The amount of strain or stress was transformed into an electrical quantity proportional to it by using an unbonded-type wire strain or stress gauge.

The circuit for detection of stress, strain, and tan δ is shown in Fig. 2.2. The circuit relationships are as follows: the output

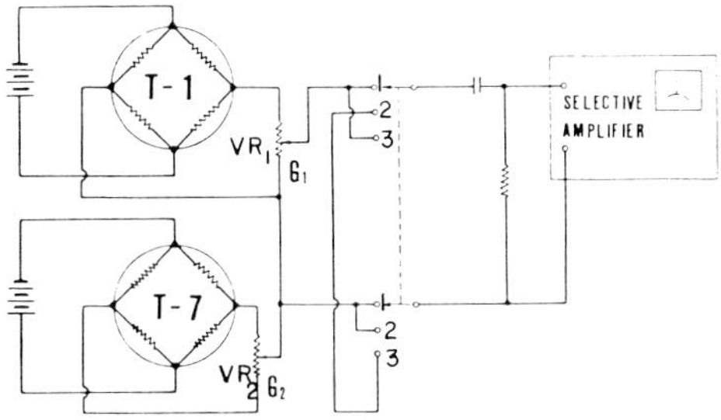

Fig. 2.2. Circuit for detection of stress, strain and tan δ.

indicated on a voltmeter is amplified with a selective amplifier through the dividers VR2 and VR1 respectively. Turning switch S to 2 first and regulating the divide ratio of VR2, the indication of the vacuum voltmeter is adjusted to 1. The divide ratio at this time is assumed to be G_2. At the same time, the switch is turned to 1 and the indication of the voltmeter is adjusted to 1 by means of VR1. The divide ratio at this time is assumed to be G_1. Turning the switch to 3 next, the subtraction of the vectors representing stress and strain is made, and the absolute value of the vector is indicated on the voltmeter, as seen in Fig. 2.2. The corresponding vector diagram is shown in Fig. 2.3. If the phase angle δ is small, $2 \sin \delta/2$

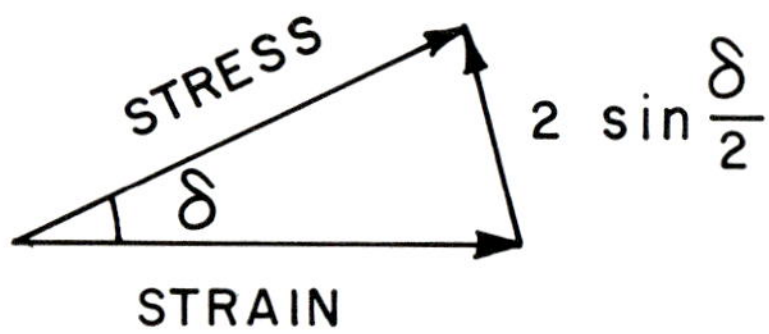

Fig. 2.3. Vector relationship among stress-strain and phase angle δ.

corresponds to $\tan \delta$. By transforming the scale of voltmeter into $\tan \delta$ beforehand, $\tan \delta$ is read off directly by the above process.

If the divide ratios G_1 and G_2 are recorded, the dynamic modulus E' and the loss modulus E'' can be evaluated simultaneously in the following relationships. When the sinusoidal tensile strain with amplitude ε_o is applied to one end of a sample of length L and cross-sectional area A, the sinusoidal stress amplitude σ_o is generated at the other end. The absolute value of the complex modulus $|E^*|$ of this sample will be as follows:

$$|E^*| = \frac{\sigma_o}{\varepsilon_o} \frac{L}{A}. \qquad (2.1)$$

Assuming the inputs to strain gauges T_1 and T_2 are E_1 and E_2 respectively, and the output signals are e_1 and e_2 respectively, the amplitudes of stress and strain will be given by

$$\sigma_o = \frac{e_1}{E_1 f_1},$$

$$\varepsilon_o = \frac{e_2}{E_2 f_2}. \qquad (2.2)$$

f_1 and f_2 are the calibration factors of transducers T_1 and T_2. From equations (2.1) and (2.2) we can obtain

$$|E^*| = \frac{f_2 e_1 E_2 L}{f_1 e_2 E_1 A}. \tag{2.3}$$

Under the conditions in which tan δ has been directly read, $e_1 G_1$ and $e_2 G_2$ can be set equal. Thus, equation (2.3) is transformed into

$$|E^*| = \frac{f_2 E_2 L}{F_1 E_1 A} \frac{G_2}{G_1}. \tag{2.4}$$

Since the quantities other than G_2/G_1 in this equation are known, E' and E" can be calculated from the following equations by using the values of G_1, G_2, and tan δ:

$$E' = |E^*| \cos \delta, \tag{2.5}$$

$$E'' = |E^*| \sin \delta = E' \tan \delta. \tag{2.6}$$

Operationally, the dynamic moduli are calculated as follows:

$$|E^*| = 2.0 \times \frac{1}{AD} \times 10^9 \times \frac{L}{S} \text{ dyn/cm}^2, \tag{2.7}$$

where L is the sample length (cm), S is the sample cross section (cm^2), and A is an instrumental constant obtained from the value of the amplitude factor when measuring tan δ. D is the value of the dynamic force read from the dial when measuring tan δ. Since tan δ is measured directly, the dynamic modulus E' and the loss modulus can be obtained by using equations (2.5) and (2.6). The oscillating load ΔF and displacement ΔL are

$$\Delta F = \frac{N}{\phi} \times 10^7 \text{ dyn}, \tag{2.8}$$

$$\Delta L = 5.0 \times A \times N \times 10^{-3} \text{ cm}, \tag{2.9}$$

where ϕ is the value of the dynamic force from the dial, and N is a constant depending on the range of tan δ. A is obtained from the value of the amplitude factor [26].

The standard Rheovibron consists of three separate units: a base frame which supports the electromechanical drive unit, the stress and strain transducer, the temperature-control chamber, and the necessary mechanical hardware to grip the sample and apply initial strain; a table-top console which contains the amplifier and oscillator chassis, the tan δ meter, and the main controls for the instrument; and a small chassis which contains the power supply for the electronics. A schematic diagram of the instrument is shown in Fig. 2.4, and its actual appearance in Fig. 2.5.

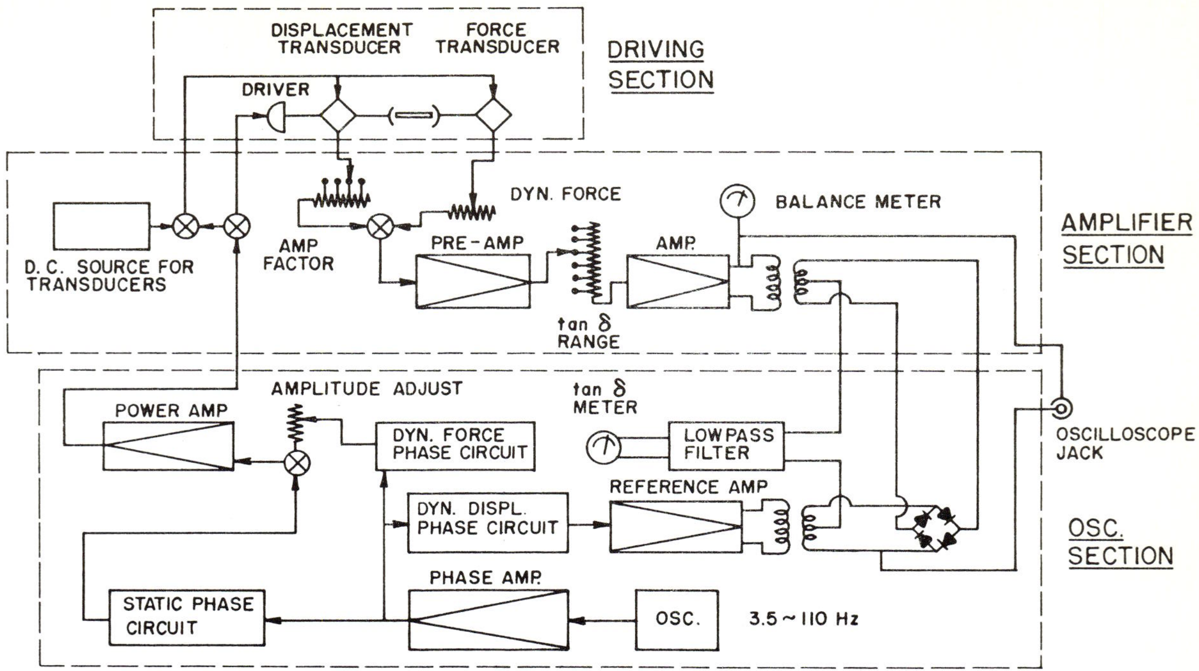

Fig. 2.4. Block diagram of Rheovibron.

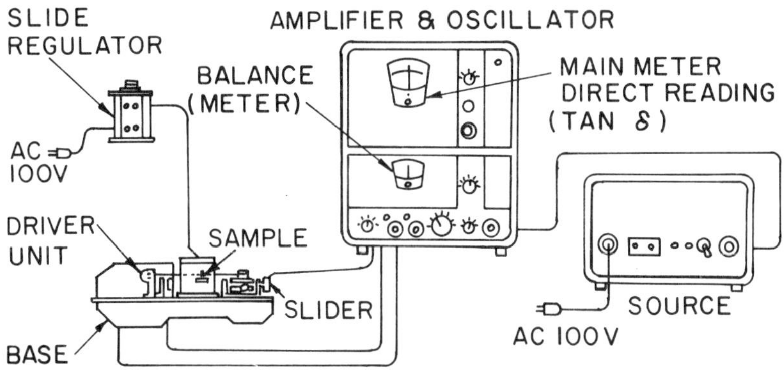

Fig. 2.5. Drawing of Rheovibron.

The features of Rheovibron are: direct reading of the dynamic loss tangent (tan δ) over a wide temperature range from -150°C to +300°C, both temperature and frequency scan on one instrument, and a wide range of measurement of the dynamic modulus from 10^6 to 10^{12} dyne/cm^2 and of tan δ from 0.001 to 1.7. Sample dimensions are: length 2 - 6 cm; cross-section 0.05 cm^2 or less for films and fibers, and 0.2 cm^2 for rubbers.

One of the most used instruments for measurements of dynamic mechanical properties of polymers is the standard Rheovibron visco-elastometer DDV-II. A number of improvements in this instrument have been reported by users. Massa [27] has proposed a mechanical model and modified the drive expressions for the moduli to take into account system compliance, sample yielding within the tensile grips, and system inertia. These effects require a correction to the raw data of 20 - 30% for glassy polymer samples in the tensile mode. The correction factor varies substantially with temperature, frequency, sample width, and thickness, and from one polymer to another. In addition to the standard tensile grips, a new sample holder has been introduced which adapts the Rheovibron to the flexural mode. A shearing mode of operation has been introduced by Erhardt [28] for investigation of the rheological properties of styrene-ethylene oxide block copolymers.

These modifications of grips have been extended to the compression mode [29] and the bending mode [30] for measurements of anisotropic dynamic mechanical properties. In addition, the Model DDV-III-B has been introduced, with larger stress capability (5 kg) and a hydraulic driving system. Using this model, a new type of instrument has been developed for measurement of the complex piezoelectric stress constant e* and the complex tensile modulus E* with devices installed in the same equipment. The principle of measurement combines the merits of the null-detection method and the direct-reading method [31]. The apparatus consists of two transducers, a hydraulic piston-servoactuator, a change amplifier, and a main amplifier. The specimen, in the form of a sheet, is subjected to a sinusoidal strain of constant amplitude by the hydraulic piston-servoactuator. Both ends are clamped in chucks. One of these chucks is forced into sinusoidal motion by oil pressure, and its longitudinal displacement is detected by a strain gauge; the other is connected to a stress gauge.

The polarization charge on the sample surface generated by the sinusoidal strain is detected with a charge amplifier. From the strain, stress, and polarization signals, the complex tensile modulus and the complex piezoelectric stress constant under the same conditions of temperature and frequency are evaluated. Detailed studies on piezoelectric properties of oriented films of poly(γ-methyl ϕ-glutamate) using this instrument have been reported [32]. The piezoelectric properties of this PMDG film are related to molecular motion and the α-helix. The piezoelectric properties of some bipolymers, such as silk and collagen, and some synthetic polymers, such as polyvinylidene fluoride, have been found to be comparable to those of inorganic piezoelectric crystals, such as quartz and Rochelle salt [33].

The automated Rheovibron was originally developed by Monsanto Co. [34], and the automation system for the Rheovibron has been commercialized by IMASS, Inc. This automatic instrument involves primarily a measurement system and an automatic tension system. The key component in the measurement system is a modified Princeton Applied Research Model 129A two-phase vector lock-in amplifier. This amplifier is used in the vector-voltmeter mode to measure the amplitude and phase of both the strain (T-7) and the stress (T-1) gauge outputs. The information of the vector voltmeter is connected to the data-logging system. A digital thermocouple thermometer and a sample-length measurement circuit are used for length- and temperature-change information for logging by the data system.

A tensioning system has been built to automatically control the tension on the sample. There are two modes of operation under control: the constant-peak tension mode and the automatic tension mode [34]. Applying these modes, the measurement accuracy and the reproducibility are improved. The complete automation system consists of temperature programming, phase-angle measuring equipment, a hard-wired programmer and data logger, a servo system for controlling the tension on the sample, interfacing for the ASR-33 teletype unit, and computer software for computer programming in Fortran IV.

3. TORSION PENDULUM

A torsion pendulum is the simplest instrument for measuring the dynamic shear modulus and mechanical damping of a polymer. The frequency range of operation is from 0.01 Hz to 50 Hz. This is a free-vibration instrument. The specimen is a cylindrical rod or a rectangular block; one end is rigidly clamped, and the other is attached to an inertia disk, which is free to oscillate. The oscillations twist and untwist the specimen. The time required for one complete oscillation is the period P. The successive amplitudes A_i will decrease with time because of the damping, which gradually converts the elastic energy of the system into heat. Instruments based on the torsion pendulum have been described and used by many investigators [5-9]. A number of commercial variants are now on the market, and ASTM standard D2236-64T describes the procedure to be used.

A simple schematic diagram of a torsion pendulum is shown in Fig. 2.6. The inertia disk is mounted directly under the specimen, as in (a), or a counterweight is used for balancing the inertia disk, which carries the upper end of the specimen, as in (b). The system is set in motion and undergoes damped sinusoidal torsional oscillations. The torsional displacement-time curve is shown in Fig. 2.7. The method of recording the displacement varies from one type of apparatus to another, but frequently it is recorded automatically. From the displacement-time curve, one can determine both the dynamic shear or torsion modulus G' and the logarithmic decrement Δ. The shear modulus is given by

$$G' = (8\pi IL/\gamma^4)(1/P^2) \tag{2.10}$$

for a cylindrical specimen, and by

$$G' = (64\pi^2 IL/ubh^3)(1/P^2) \tag{2.11}$$

for a rectangular specimen. In these equations, L = length of specimen,

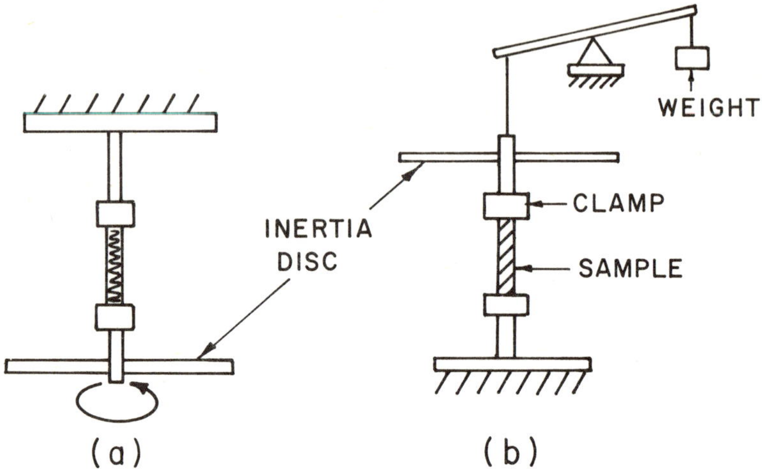

Fig. 2.6. Simple diagram of torsion pendulum.

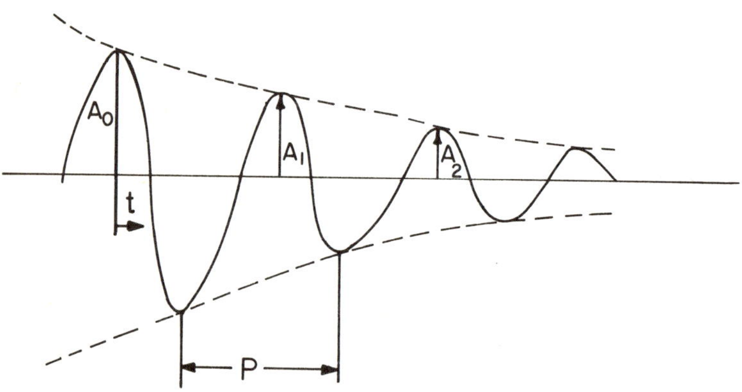

Fig. 2.7. Torsional displacement-time curve.

I = moment of inertia of the inertial member, b = specimen width, h = specimen thickness, γ = specimen radius, u = shape factor depending on the ratio of width to thickness, and P = period of oscillation in seconds.

The logrithmic decrement Δ is calculated from the logarithm (to the base e) of the ratio of the amplitudes of two successive oscillations:

$$\Delta = \ln \frac{A_1}{A_2} = \ln \frac{A_2}{A_3} = \ln \frac{A_n - 1}{A_n}, \quad (2.12)$$

Where A_1 is the amplitude of the first oscillation and A_2 is the amplitude of the second oscillation, as shown in Fig. 2.7. It is related to the dissipation factor tan δ and the loss shear modulus by

$$\Delta = \pi \tan \delta, \quad (2.13)$$

$$G''/G' = \tan \delta. \quad (2.14)$$

These equations for the shear modulus and damping are accurate only if there is no tension on the specimen. Some torsion pendulums are designed to study polymers under stress [35].

4. TORSIONAL BRAID ANALYSIS

Torsional braid analysis (TBA) is an extension of the torsion-pendulum method for examining the dynamic mechanical properties of materials. It was developed by J. K. Gillham [36], and his extensive studies on this subject can been found in more than 30 publications [13,14,25,36,37]. This method is the only one available for measuring properties of solids or melts with less than 100 mg of polymer per experiment using the torsion pendulum. The technique differs from the simple torsion-pendulum one mainly in that the sample specimen is prepared by impregnating a multifilament glass braid substrate with a solution of the material which is to be investigated. Changes in the mechanical properties of the composite are attributed to changes in the polymer. An advantage of using a supported-polymer technique is that specimen fabrication is generally simplified to deposition of the material from a melt or from solvent onto the substrate. Further, by the use of a support, a material can be investigated through various regions of the mechanical spectrum, including those in which the material may not be capable of supporting even its own weight. This provides an opportunity for studying resin-forming reactions and high-temperature reactions of thermoplastic polymers, both of which often start in the liquid state and proceed through gel, rubbery, and glassy states.

In addition, Boyer's [38] comprehensive review on mechanical motion in amorphous and semi-crystalline polymers describes how TBA provides information for the oligomeric range of molecular weights as well as the region above T_g, or even the liquid region above T_m, by using pendulums with different moments of inertia.

The shear modulus G' can be obtained by using equation (2.10); however, since the experimental system contains a heterogeneous supporting member, this equation is not strictly applicable. In torsional braid analysis the relative rigidity parameter $1/P^2$, where P is the period of oscillation, is used as a measure of the modulus. Of major significance is the fact that the relative rigidity is independent of the sample geometry, thus eliminating any problems which may arise in interpreting these quantities. The shear modulus can be calculated from the period P—the shorter the period, the greater the modulus. The damping factor is obtained from equation (2.12). However, the method used to calculate the mechanical damping is based on the fact that for a damped sine wave the number n of oscillations which occur between two fixed but arbitrary boundary conditions (e.g., A_o/A_n = 30/1), is inversely proportional to the logarithmic decrement. The resulting measure (1/n) of the mechanical damping is called the mechanical-damping index.

The TBA apparatus consists of a cylindrical glass column constructed so as to enclose the test specimen and inertial mass in a controlled atmosphere throughout the experiment. The mechanical oscillation is initiated at the top of the apparatus, while a transducer is sensing the oscillations electrically to provide a graphical record on a strip-chart recorder. The instrumentation has been improved during the past ten years, and Gillham [36] has automated his TBA equipment to the point that temperature programming is completely controlled in any desired logical sequence, while an X-Y-Y' recorder plots out, as a function of temperature at ≃1 Hz, log decrement and log (period^{-2}) the latter being a measure of relative modulus rigidity.

TBA has been utilized with differential thermal analysis (DTA) and thermogravimetric analysis (TGA) for studies on cellulose triacetate, poly(vinyl alcohol), poly(ethylene terephthalate), polyacrylonitrile, polycaprolactam (nylon 6), and many other polymers. The results showed a high degree of correlation between the thermomechanical and thermal spectra of the polymers [39].

5. VIBRATING REED

The vibrating reed [12,16-18], or cantilever, is a simple and widely used technique for measuring Young's modulus and internal friction or damping. A small specimen in the form of a plastic strip or reed is clamped at one end in an electromagnetic vibrator which is driven by a variable-frequency oscillator. As the frequency of the vibrations is changed, the natural resonance of the reed will be reached, and the amplitude of the free end of the reed will go through a maximum. The amplitude of vibration of the specimen is measured as a function of frequency at a fixed temperature.

Figure 2.8 shows a simple diagram of the vibrating-reed apparatus. The resonance curve, which shows the change of amplitude of the free

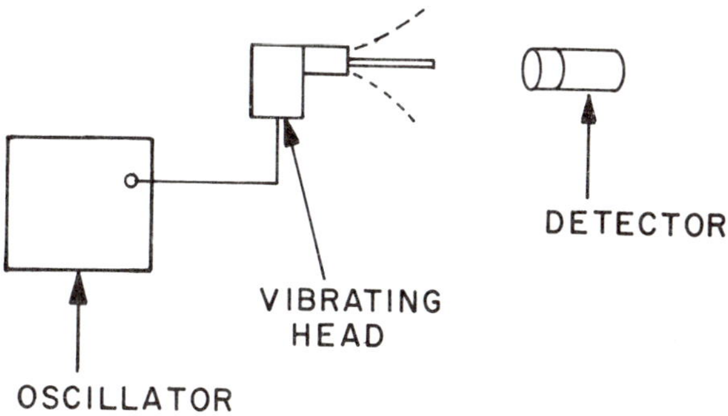

Fig. 2.8. Schematic diagram of the vibrating-reed technique.

end of the reed with frequency of exciting vibration, is shown in Fig. 2.9. In order to measure the motion of the specimen, a light source and a photocell with an oscilloscope or a micrometer eyepiece on a cathetometer are used [16,40]. It is a simple arrangement, but the dynamic viscoelastic parameters can be obtained. The vibroscope is an extension of the vibrating-reed technique for measuring the size of a filament by vibration [41].

Young's modulus E in dyn/cm^2 is calculated from

$$E = \frac{38.24\rho L^4}{h^2} f_r, \qquad (2.15)$$

where ρ is the density of plastic specimen in g/cc, L is the free length of reed in centimeters, h is the thickness of the specimen in centimeters, and f_r is the resonance frequency in hertz.

A damping term may be calculated from the sharpness of the resonance peak,

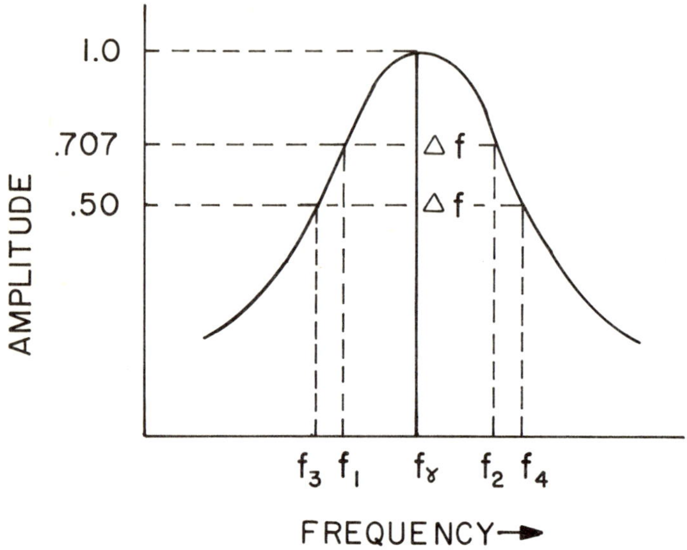

Fig. 2.9. Resonance curve of free end.

$$\tan \delta = \frac{\Delta f}{f_r}, \tag{2.16}$$

where Δf is the bandwidth, defined as the difference between those frequencies for which the amplitude of vibration has $1/\sqrt{2}$ times its maximum value. (Sometimes it is taken between frequencies at which the amplitude is one-half the maximum.)

6. MECHANICAL SPECTROMETER, DMA, AND B&K APPARATUS

(1) Mechanical spectrometer. The mechanical spectrometer is a rotational rheometer for measurement of the basic rheological properties, such as shear viscosity, normal stresses, dynamic moduli, and melt elasticity in fluids and solids. The Weissenberg Rheogoniometer [42] by Sangamo Controls Ltd., England, and the Rheometrics Mechanical Spectrometer by Rheometric, Inc. [43], U.S.A., are examples of the commercial instruments.

This apparatus has been used extensively to study polymer melts [44-51] by using various geometrics such as eccentric rotating disks, cone and plate, parallel plates, and concentric cylinders [52-59]. Sample specimens of polymer melts, rubbers, and elastomers (about 1.5 grams) are placed between the two disks, as seen in Fig. 2.10. The

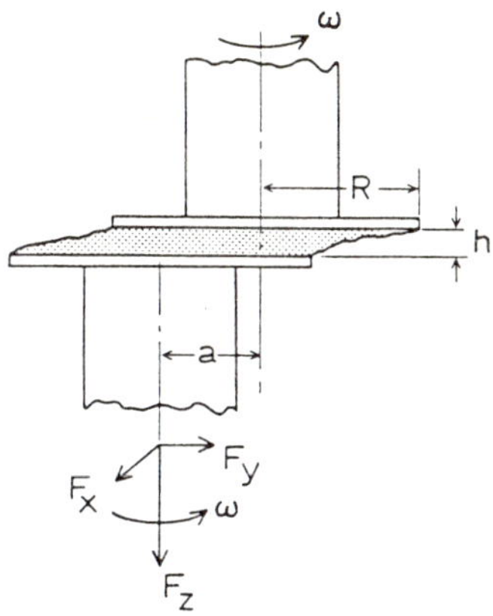

Fig. 2.10. Flow between eccentric rotating disks.

instrument deforms the specimen rotationally and measures the stress responses with a unique transducer, which is capable of simultaneous detection of three orthogonal forces (F_x, F_y, and F_z) and the torque about the z axis. This solid-state unit continuously monitors three forces: the normal force and the torque in cone-and-plate flow, the biaxial stress relaxation in a solid rod, or the three orthogonal forces generated by the flow between eccentric rotating disks.

The dynamic moduli can be calculated from these detected stresses. For instance, in the case of the eccentric rotating disks, two circular disks rotate about parallel axes that have been displaced a small amount a, as seen in Fig. 2.10. The sample, of thickness h, is confined between the disks and, due to the eccentricity, is forced to go through a shear cycle with each revolution. When the disk rotates 90°, elements of the sample are subjected to shear deformation. This shear strain results in a shear stress or force on the surface of each disk. F_y is the elastic component of the force; it tends to return the sample to its undeformed state. It is a measure of the energy stored, and is related to the elastic modulus G' by a factor of the stress divided by the strain:

$$G' = \frac{F_y/\pi R^2}{a/h}. \tag{2.17}$$

Materials also dissipate energy. Thus, there will be a viscous component of the shear force, F_x. It is a measure of energy loss and is related to the viscous or loss modulus G" by a factor of the stress divided by the strain:

$$G'' = \frac{F_x/\pi R^2}{a/h}. \tag{2.18}$$

From the stresses F_z and F_y with h and a, the normal stress difference can be obtained.

The basic mechanical spectrometer unit consists of a transducer with full scale equal to 20 g to 10 kg of orthogonal force (x, y, and z) and 200 g cm to 100 kg cm of torque. The fundamental drive unit consists of this transducer plus five additional units: a precision air bearing; a signal conditioner; a sample thermocouple with digital readout; a unit for x, y, and z translation; and a constant-angular-velocity drive from 0.001 to 200 rad/sec with continuous and discrete speed selection.

Efficient environmental chambers allow quick and accurate measurements from -120°C to 320°C for solids, melts, and fluids, controlled by either manual or programmable temperature controllers. Dynamic oscillatory or vibrational motions can be provided through a low-frequency sinusoidal function generator with a useful frequency range of 0.001 to 30 Hz. An automatic-printout option for all rheological test data, as well as paper-punch units and X-Y plotting systems, can conveniently be added to the automated components. A sensitive transducer unit for very fluid systems (water-based polymers, biological fluids, emulsions, and paint solutions) is interchangeable with the standard transducer. It measures torque about the z-axis from 0.2 to 100 g cm and a normal force from 2.0 to 1000 g full scale [43].

In order to measure the dynamic properties of materials, from low-viscosity fluids to reinforced solid composites, this spectrometer uses various test fixtures in addition to the transducer. Figure 2.11 shows representative test fixtures for solution, melt, rubber, and solid specimens.

The cone-and-plate mode is used frequently for viscosity and normal-stress characterization of low- to medium-viscosity polymer melts. These measurements are often related to blow-molding behavior. The eccentric-rotating-disk modes have been used particularly in the area of crosslinking reactions typical of epoxies, phenolics, rubber, and sheet molding compounds. Macosko's extensive investigations into the relationships between these rheological parameters and polymer processing [47,55-59] have developed useful applications.

The EMR Schulmberger Frequency Response Analyzer (Weston Instruments Inc., Model 1172) is used for dynamic-property measurements in conjunction with the Weissenberg Rheogoniometer and the Wallace Shawbury

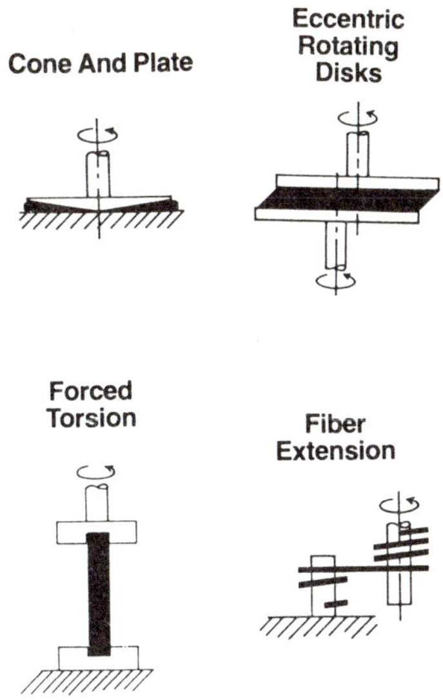

Fig. 2.11. Representative test fixtures of mechanical spectrometer.

sinusoidal-strain machine [60]. This analyzer applies a frequency of 0.001 Hz to 10 kHz to the driving transducer. The signal from the receiving transducer goes back to the analyzer, which displays the dynamic response. Measured properties are the complex viscosity and the dynamic moduli.

(2) Dynamic Mechanical Analyzer (DMA). The newest commercial mechanical spectroscope is the DuPont 980 Dynamic Mechanical Analyzer (DMA). This instrument measures the resonant frequency and energy dissipation without tension on a wide range of samples as a function of temperature (or time) [61]. The resonant frequency is related to the modulus by the sample geometry. Energy dissipation relates to such properties as impact resistance, brittleness, and noise abatement.

The analyzer contains the mechanical oscillator, driver, and digital display portions of the DMA system. The heart of the instrument is a set of arms mounted on low-friction flexure pivots which have a system resonant frequency without sample of less than 3 Hz. One of these arms (the driven sample arm) acts as a compliant axis, while the other (counterweighted) arm acts as the axis of physical support. This arrangement, which is converted to a compound resonance system by the sample, allows the sample through its modulus to contribute as much of the overall restoring force as possible, while maintaining overall structural integrity.

A sample of known dimensions is clamped perpendicularly between the two arms. The compound resonance system (sample and arms) is then driven by an electromechanical transducer to maintain constant-amplitude flexural oscillation at the sample's natural resonant frequency. In operation, the active arm is rotated slightly around its pivot, generating a few tenths of a millimeter deflection at the sample. This puts the sample in flexural stress. When the displacing force is released, the deflection energy stored in the sample causes it to go into resonant oscillation. The frequency and amplitude of this oscillation are sensed by a linear variable differential (LVDT). An LVDT sensor monitors the movement of the sample system through the driven arm with a 1-second gate. It determines both the number of oscillations per second (frequency) and the amplitude of the oscillations for the sample system. The frequency of the resonance system is fed from the LVDT to both a direct light-emitting-diode (LED) read out on the front panel of the instrument and one axis of the X-Y,Y' recorder. The frequency readout is a linear scale in hertz. The sample modulus is obtained from the frequency data by a formula involving the sample geometry. The modulus equation is [61]

$$E = \frac{4\pi^2 f_o^2 J_o - K}{2H(\frac{D}{2} + L_V)^2} \quad (\frac{D}{B})^3, \qquad (2.19)$$

where E is Young's modulus in dyn/cm², f_o is the DMA frequency in Hz, J_o is the moment of inertia of the arm in g cm², K is the spring constant of pivot in dyn/cm rad^{-1}, L_V is the clamping distance in cm, H is the sample width in cm, B is the sample thickness in cm, D is the sample length in cm. The energy dissipation of the sample is directly measured by the amount of power necessary to maintain constant amplitude. This power is fed to another LED readout on the front panel of the 980 analyzer and to the Y-axis of the X-Y,Y' recorder. The sample damping (energy dissipation) varies by several decades during a transition, and so it is displayed on a logarithmic scale in decibel (dB). The temperature is programmed from -150°C to 500°C at various heating rates of 0.5 to 20°C/min; the sample is 0.02 to 1.6 mm thick, 0.02 to 13 mm wide, and 10 to 25 mm long. The modulus range is 0 to 10^{11} dyn/cm², so that metals, polymers, and other materials can be studied. The frequency varies as the temperature increases and the modulus decreases; the variation is generally in the range from 150 to 3.5 Hz. In the specific instance of linear polyethylene, the variation is from 20 to 3.5 Hz. The output information from the unit can be fed to computers for modulus calculations.

(3) B&K complex-modulus apparatus. The complex-modulus apparatus has been developed to determine the complex modulus of elasticity and the loss factor of solid materials. In the development of this apparatus, there was special regard for the measurement of plastic materials in the vital frequency range of 10 to 10,000 Hz.

A block diagram of the measuring arrangement is shown in Fig. 2.12. This instrument consists of four units: the beat-frequency oscillator

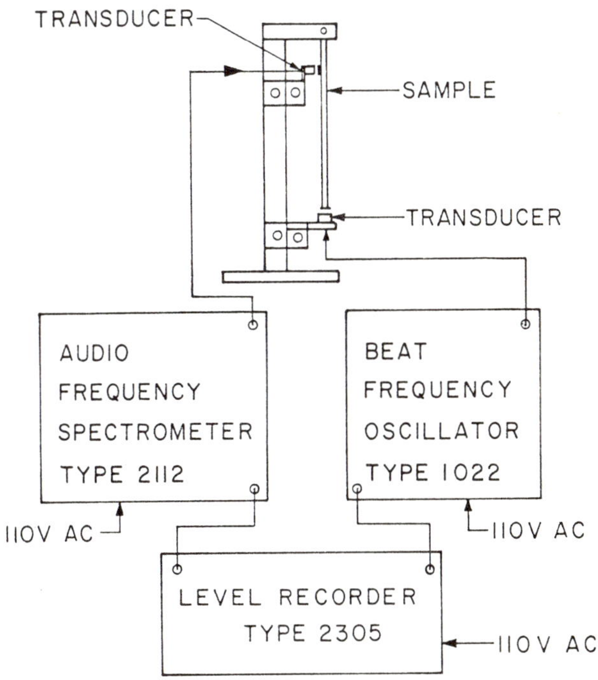

Fig. 2.12. Block diagram of acoustic spectrometer.

(Type 1022), the audio-frequency spectrometer (Type 2112), the complex-modulus apparatus (Type 3930), and the level recorder (Type 2305). These units are made by the Brüel and Kjaev Company [63].

The sample bar is clamped at one or both ends in the complex-modulus apparatus. Excitation and detection are accomplished with magnetic transducers. Small disks of ferromagnetic material are glued to the sample bar in front of each transducer. The beat-frequency oscillator excites the sample bar at frequencies from 20 Hz to 20 kHz. The displacement of the bar is detected by a transducer and the audio-frequency spectrometer. The level recorder draws frequency-response curves automatically on the chart. Usually, four to six resonance modes are obtained on this curve. It is necessary to determine the ratios between different resonance frequencies.

Knowing the resonance frequency of the beam and its dimensions, the modulus of elasticity can be calculated from the following equation:

$$E' = \frac{48\pi^2\rho\ell}{d^2}\,\frac{f_r}{K_y}, \tag{2.20}$$

where E' is the dynamic modulus in dyn/cm^2, ρ is the density of the specimen in g/cc, ℓ is the free length of the beam, d is the thickness of the specimen, k_y is the tabulated eigenvalue for the yth mode [64], and f_r is the resonance frequency of the yth mode in hertz.

The loss factor (damping factor) is determined from the sharpness of resonance curves or the decay rate of a resonance when the excitation is stopped. The decay rate is obtained from the slope of the displacement-versus-time plot on the level-recorder chart.

The sample bars should be cut with rectangular cross section (max. 12 X 12 mm) and length 7.5 to 22 cm. However, fibers can be examined by using the modified complex-modulus apparatus containing the mechanical system of the fiber and the mass. The mass is excited into vertical oscillation by a shaker.

With the environmental chamber, this apparatus is designed for use over a very great temperature range, from extremely low temperatures (-150°C) up to about 250°C. It is thus possible to measure the properties of the material as a function of temperature as well as frequency.

7. SONIC TECHNIQUE

The sonic technique has been employed to determine the dynamic Young's modulus of a wide range of materials (e.g., metal rods [65], concrete beams [66], paper [67], and fibers [68]). The dynamic modulus E of a solid material is calculated from the acoustic velocity V by the relationship [69]

$$E = \rho V^2, \tag{2.21}$$

where ρ is the solid density. This velocity is independent of sample geometry [65]. In practice, a sonic pulse is transmitted along a length of solid, and the flight time between transmitting and receiving transducers is measured. From a knowledge of the instrument delay time and sample length, the sonic velocity can be computed and used in equation (2.21).

The PPM-5 (pulse propagation meter) dynamic-modulus tester developed by the Henry H. Morgan Company has been commonly used in the fiber field. Sonic velocity is measured in the PPM-5 by measuring the elapsed time for a longitudinal mechanical pulse to travel a given distance through the sample. Continuous pulse transmission and timing provide continuous readout on a strip-chart recorder. The sonic velocity is calculated from length and time measurements. The tester has also been modified to measure flight times over specified lengths on test specimens [68]. The transducers of the PPM-5 were mounted vertically on the twist-counter frame, directly beneath and in contact with the test specimen, whose length was 20 cm or 40 cm. The flight times of the strain pulses, which were generated by the instrument, were measured on a Tektronix oscilloscope. A block diagram of this modified instrument is shown in Fig. 2.13. The pulse generator sends a high-voltage spike pulse to the transmitting transducer (60 pulses per second), while the pulse generator simultaneously sends a trigger pulse to the oscilloscope to begin the sweep. A longitudinal mechanical pulse, which is transmitted along the yarn, excites the receiving transducer and simultaneously amplifies the waveform while displaying it on the oscilloscope. The time base of the oscilloscope is then used to read the time between the start of the trace and the beginning of the transducer waveform, as shown in Fig. 2.14. The pulse transit time and the distance between the transducers are factors in determining sonic velocity. The transducers are piezoelectric ceramic ones, covering a wide temperature range. The pulse-generated displacement of the transmitting transducer is about 0.025 mm. Therefore, a very small strain amplitude is transmitted to the yarn at a frequency of around 10 kHz.

The PPM-5 instrument provides a continuous readout on a strip-chart recorder for measuring the elapsed time with a scanning transducer. But, due to poor contact between sample and transducer, the scanning system is not adequate for determining transit times on rough-surface yarn and twisted yarn. Therefore, this instrument has

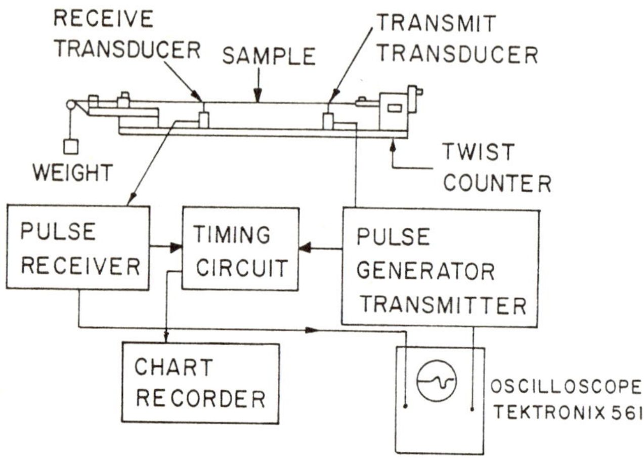

Fig. 2.13. Block diagram of pulse-propagation apparatus.

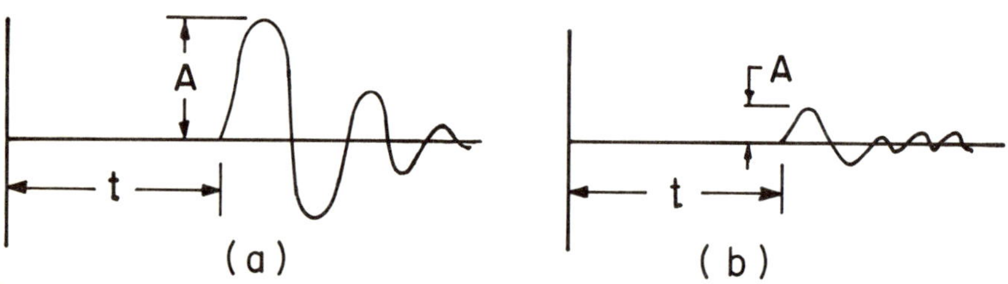

Fig. 2.14. (a) Good transducer contact: high signal amplitude A; (b) poor transducer contact: lower signal amplitude A; t = flight time.

been modified by using an oscilloscope instead of a strip-chart recorder and scanner.

Two important parameters in determining the accuracy of sonic-velocity measurements are time delay and transducer contact. A short delay is unavoidably introduced by the circuit that triggers the transmitter. This delay includes the time needed to transmit the electrical signal. The delay time may also be thought of as the transit time if the transducer separation were zero. Time-measuring transducers should be mounted in such a manner that the distance between them can be varied. The transit time is then plotted against the distance between transducers. The zero-intercept value of the transit time is equal to the delay, and this quantity must be subtracted from all readings to obtain true values of the transit time. Even though the delay time is constant over a wide range of materials, it is dependent upon temperatue and the type of transducer employed. Therefore, the delay time of each transducer is determined by calibration before further experimentation. The delay time in this system is ≈5 to 8 μsec. Transducer contact affects the signal amplitude, and poor contact between the sample and the transducers produces a low signal amplitude on the oscilloscope, as indicated in Fig. 2.14.

A standard sample, fused quartz, is used to calibrate the instrument. The sonic velocity of fused quartz is given as 5760 m/sec for standardization purposes. The sonic velocity determined by this instrument was 5809 meters per second, a difference of 0.85%.

REFERENCES

1 J. D. Ferry, Viscoelastic Properties of Polymers, 2nd ed., Wiley, New York, 1969.
2 L. E. Nielsen, Mechanical Properties of Polymers and Composites, Marcel Dekker, New York, 1974.
3 I. M. Ward, Mechanical Properties of Solid Polymers, Wiley, New York, 1971.
4 J. H. Dillon and S. D. Gehman, India Rubber World, 115 (Oct. 1946) 61.
5 L. E. Nielsen, Rev. Sci. Instrum., 22 (1951) 690.
6 K. H. Illers and Jenckel, J. Polym. Sci., 41 (1959) 528.
7 J. Koppelmann, Kolloid-Z. Z.-Polym., 144 (1955) 12.
8 D. J. Plazek, M. N. Vrancken, and J. W. Berge, Trans. Soc. Rheol., 2 (1958) 39.
9 N. Tokita, J. Polym. Sci., 20 (1956) 515.
10 F. Rodrigues, R. A. Van Brederode, and G. G. Cocks, J. Appl. Polym. Sci., 12 (1968) 2415.
11 C. J. Nederveen and C. W. Van der Wal, Rheologica Acta, 6 (1967) 316.
12 K. Fujino, H. Kawai, and M. Horio, Textile Res. J., 25 (1955) 722.
13 J. K. Gillham, Polym. Eng. Sci., 7 (1967) 225.
14 J. K. Gillham and M. B. Roller, Polym. Eng. Sci., 11 (1971) 295.
15 J. W. Ballou and J. C. Smith, J. Appl. Phys., 20 (1949) 493.
16 M. Horio, S. Onogi, C. Nakayama, and K. Yamamoto, J. Appl. Phys., 22 (1951) 966.
17 R. Buchdahl, R. J. Morgan, and L. E. Nielsen, Rev. Sci. Instrum., 41 (1970) 1342.
18 A. N. Gent, Br. J. Appl. Phys., 11 (1960) 165.
19 R. S. Marvin, E. R. Fitzgerald, and J. D. Ferry, J. Appl. Phys., 21 (1950) 197.
20 P. Lord and R. E. Wetton, J. Sci. Instrum., 38 (1961) 385.
21 R. N. Shroff, D. I. Livingston, and G.S. Fielding-Russell, Rubber Chem. Technol., 43 (1970) 1491.

22 M. Yoshino and M. Takayanagi, J. Jap. Soc. Test. Mater., 8 (April 1959) 330.
23 M. Takayanagi, Viscoelastic Properties of Crystalline Polymers, Mem. of the Fac. of Eng., Kyushu Univ., Vol. 23 (1963), No. 1.
24 R. F. Boyer, in Polymeric Materials: Relationship between Structur and Mechanical Behavior (E. Baer and S. V. Radcliffe, Eds.), Am. Soc. Metals, Ohio, 1974, Chap. 6, pp. 277-368.
25 J. K. Gillham, S. J. Stadnicki, and Y. Hazony, J. Appl. Polym. Sci., 21 (1977) 401.
26 Vibron, Direct Reading Dynamic Viscoelastometer Model DDV-II, Instruction Manual 17, Toyo Measuring Instruments Co., Tokyo, 1966.
27 D. J. Massa, J. Appl. Phys., 44 (1973) 2595.
28 P. F. Erhardt, J. J. O'Malley, and R. G. Crystal, in Block Copolymers (S. L. Aggarwal, Ed.), Plenum, New York, 1970, pp. 195-213.
29 T. Murayama, J. Appl. Polym. Sci., 20 (1976) 2593.
30 T. Murayama, J. Appl. Polym. Sci. (in press).
31 K. Koga, T. Kajiyama, and M. Takayanagi, J. Phys. E: Sci. Instrum. 8 (1975) 299.
32 K. Koga, T. Kajiyama, and M. Takayanagi, J. Polym. Sci., Phys. Ed., 14 (1976) 401.
33 E. Fukada, Progr. Polym. Sci., 2 (1971) 329.
34 A. S. Kenyon, W. A. Grote, D. A. Wallace, and M. C. Rayford, ACS Polymer Preprints, 17, #2 (Aug. 1976), 7-12.
35 P. Heydemann and A. Zosel, Acustica, 15 (1965) 49.
36 J. K. Gillham, in Techniques and Methods of Polymer Evaluation, Vol. 2 (P. E. Slade, Jr., and L. T. Jenkins, Eds.), Marcel Dekker, New York, 1970, p. 225.
37 R. F. Boyer and J. K. Gillham, Polym. Prepr., 18 (Mar. 1977), No. 1, 623.
38 R. F. Boyer, Polymer, 17 (1976) 996.
39 J. K. Gillham, Appl. Polym. Symp., No. 2 (1966), 45,49.
40 S. Strella, ASTM Bull., No. 214 (1956) 47.
41 D. J. Montgomery and W. T. Milloway, Textile Res. J., 22 (1952) 729.
42 Sangamo Controls Ltd., Weissenberg Rheogoniometer Instruction Manual, Sussex, England.
43 Rheometrics, Inc., Mechanical Spectrometer, Product Bulletin, Union, N. J., U.S.A., 1976.
44 R. L. Cerro, C. W. Macosko, and L. E. Scriven, Nat. Phys. Sci., 241 (1973) 146.
45 F. G. Mussatti and C. W. Macosko, Polym. Eng. Sci., 13 (1973) 236.
46 G. S. Kyker and T. A. Antkowiak, Rubber Chem. Technol., 47 (1974) 32.
47 C. W. Macosko and F. G. Mussatti, Polym. Eng. Sci., 13 (1973) 236; Rheologica Acta, 12 (1973) 289.
48 J. H. Southern and D. R. Paul, Polym. Eng. Sci., 14 (1974) 560.
49 N. Nakajima, D. W. Ward, and E. A. Collins, J. Appl. Polym. Sci., 20 (1976) 1187.
50 N. W. Taylor and E. B. Bagley, J. Polym. Sci., 13 (1975) 1133.
51 C. W. Macosko and P. J. Morse, VIIth InternationalCongress on Rheology, Gothenburg, Sweden, Proceeding Preprint 376 (Aug. 1976).
52 J. Harris and W. L. Wilkinson, Br. Soc. of Rheol. Bull. No. 1 (1969) 1.
53 K. Walters and R. A. Kemp, Rheologica Acta, 7 (1968) 1.
54 B. Warburton and S. S. Davis, Rheologica Acta, 8 (1969) 205.
55 C. W. Macosko and J. M. Starita, SPE J., 27 (1971) 38.
56 C. W. Macosko and F. C. Weissert, Am. Soc. Test. Mater., Spec. Tech. Pub. No. 553, Philadelphia, Pa., 1974.
57 W. M. Davis and C. W. Macosko, AIChE J., 20 (1974) 600.
58 E. Broyer and C. W. Macosko, SPE Tech. Papers, 21 (1975) 343.

59 W. M. Davis and C. W. Macosko, SPE Tech. Papers, 22 (1976) 408.
60 EMR Schlumberger, Frequency Response Analysis Techniques for Materials Testing, Solartron Electronic Group Ltd., Farnborough, Hampshire, England, 1976.
61 DuPont Instruments, 980 Dynamic Mechanical Analysis System (DMA), 1976.
62 R. L. Blaine and L. Woo, Polymer Preprints, 17, No. 2 (Aug. 1976) 1.
63 Bruel and Kjaer Instruments, Instructions and Applications of Complex Modulus Apparatus Type 3930, Copenhagen, Denmark, 1974.
64 J. P. Den Hartog, Mechanical Vibrations, McGraw-Hill, New York, 1956, p. 139.
65 R. C. McMaster, Nondestructive Testing Handbook, Ronald Press, New York, 1963, Sect. 43.9.
66 E. A. Whitehurst, Evaluation of Concrete Properties from Sonic Tests, ACI Monogr. No. 2, 1966, p. 3.
67 J. K. Craver and D. L. Taylor, Nondestructive Sonic Measurement of Paper Elasticity, Tappi Plastic Conference, 19-21 Oct. 1964.
68 C. F. Zorowski and T. Murayama, Textile Res. J., 37 (1967) 852.
69 J. W. Rayleigh, Theory of Sound, Dover, New York, 1945 (Orig. Edn. 1877).

CHAPTER III

MOLECULAR INTERPRETATION

1. INTRODUCTION

Dynamic mechanical properties can be determined by a number of previously discussed instruments (cf. Chap. II). Some instruments are designed for rubbers and polymer melts, whereas others are for rigid polymers or fibers. The dynamic data are obtained as a function of temperature and frequency.

The dynamic modulus and internal friction are the most basic of all mechanical properties, and their importance in any end-use application is well known. The dynamic loss modulus, or damping, is sensitive not only to many kinds of molecular motion, but also to various transitions, relaxation processes, structural heterogeneities, and the morphology of multiphase systems (crystalline polymers, polymer blends, and copolymers). Therefore, interpretations of the dynamic mechanical properties at the molecular level are of great scientific and practical importance in understanding the mechanical behavior of polymers.

Much research has been done on the relationships between molecular parameters and the dynamic mechanical behavior of polymers. The book by McCrum, Read, and Williams [1], published in 1967, gives a comprehensive review of dynamic mechanical spectroscopy in relation to the molecular structure of polymers. A number of useful generalizations and interpretations on this subject are found in texts by Nielsen [2] and Ferry [3]. Takayanagi has proposed new interpretations on the relaxation spectrum [4-6]. In recent years, Boyer also has reviewed the molecular motions in amorphous and semicrystalline polymers with respect to the anelastic spectra [7,8]. This has been the most activ area in the dynamic mechanical analysis of polymers.

2. PRIMARY AND SECONDARY TRANSITIONS

Typical dynamic mechanical properties of amorphous polymers as a function of temperature are illustrated in Fig. 3.1.

In the region where the dynamic-modulus-temperature curve has an inflection point, the internal friction (tan δ) curve goes through a maximum. This dispersion occurs in the glass transition region. The

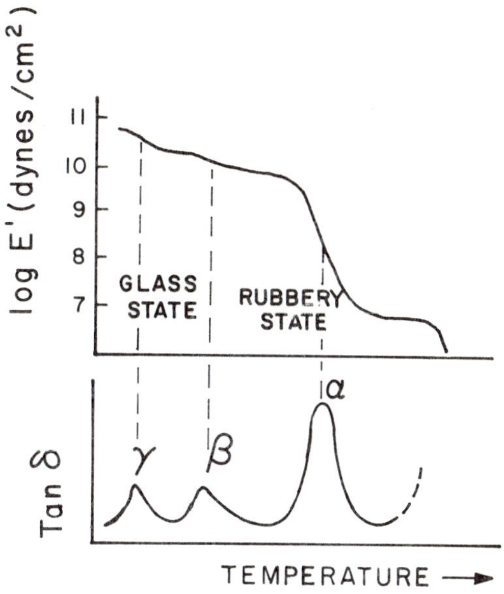

Fig. 3.1. Typical dynamic mechanical properties of amorphous polymers.

dynamic modulus E' changes from approximately 10^{10} dyn/cm^2 in the glassy state to about 10^7 dyn/cm^2 in the soft rubbery state.

In the transition region the damping is high owing to the initiation of micro-Brownian motion in molecular chains. Some of the molecular chain segments are free to move, while others are not. A frozen-in segment can store much more energy for a given deformation than can a free-to-move rubbery segment. Thus, every time a stressed, frozen-in segment becomes free to move, its excess energy is dissipated as heat. Micro-Brownian motion is concerned with the cooperative diffusional motion of main chain segments. This transition is so conspicuous that it is called the primary dispersion (the α-peak).

The loss modulus E" goes through a peak at a slightly lower temperature than does the internal friction E"/E'. The maximum heat dissipation per unit deformation occurs at the temperature where E" is maximum; this temperature at 1 Hz is very close to the value of the glass-transition temperature as determined by volume-temperature measurements.

Other relaxation transitions can be found in a glassy state on the lower-temperature side of the primary dispersion. These are called secondary dispersions and are usually designated β, γ, etc., in order of decreasing temperature. The β-dispersion, for example, has been shown by a combination of comparative NMR and dielectric-measurement studies on similar polymers to be associated with side-chain motion of the ester group [9-12]. The γ-dispersion involves motion of the methyl groups attached to the main chain. The activatio energy of these secondary dispersions is generally lower than that of the α-dispersion, since it corresponds to motion of segments that is smaller than the micro-Brownian motion of the main chain.

Figure 3.2 shows typical dynamic mechanical properties of a crystal line polymer as a function of temperature. The primary relaxation

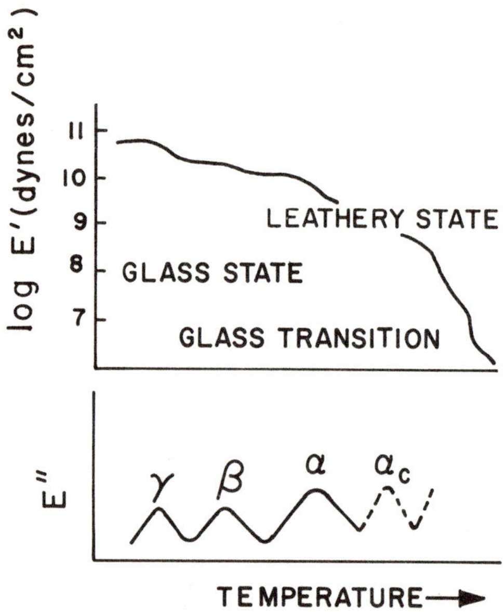

Fig. 3.2. Typical dynamic mechanical properties of crystalline polymers.

(α-peak), corresponding to the glass transition (T_g), results from the initiation of micro-Brownian motion of the amorphous chains. In this glass-transition region the dynamic modulus of the glassy state (10^{10} - 10^{11} dyn/cm²) decreases to the leathery-state modulus of around 10^9 dyn/cm², about 100 times the modulus in the rubbery region The appearance of the leathery state is caused by the micro-Brownian motion of the noncrystalline region under the structural restraint of the neighboring molecular chains in the crystalline regions. For this reason, even after crystallizing a given polymer, there is a general tendency for the α-peak to shift to higher temperature and broaden in comparison with the amorphous state.

Highly crystalline polymers (80-90% degree of crystallinity) such as high-density polyethylene and isotatic polypropylene, exhibit another dispersion between the α-dispersion and melting temperatures. This relaxation is called the α_c- or α'-dispersion. Takayanagi attributed the α_c-relaxation to molecular motion within the crystalline phase [4]. He concluded that this crystalline dispersion is due to the frictional viscosity among specific crystalline planes or molecules inside the crystals.

In the lower temperature region below the α-peak, one can observe the β-dispersion, and in some cases the γ-dispersion. The magnitude of these peaks is lower than that of the α-peak. The interpretation of the corresponding relaxation processes is more complex. However, generally the low-temperature relaxations involve motions of relatively small groups: side groups in amorphous or crystalline regions, crystalline defects, and incompatible impurities.

3. MOLECULAR MOTION AND LOSS FACTOR

Motions of long chain segments in the polymer structure have a profound effect on the loss factors (tan δ and loss modulus) of the dynamic mechanical properties. The tan δ value at the α-peak (the glass transition) is of greater magnitude than at the dispersion peaks for lower temperatures, and is accompanied by the greatest decrease in dynamic modulus with increasing temperature.

The loss factors plotted against temperature (as seen in Figs. 3.1 and 3.2) often are called relaxation or anelastic spectra. These contain several loss peaks. Analyses of these spectra are useful for determining molecular motions in combination with dielectric nuclear-magnetic-resonance (NMR) measurements. Numerous relaxation spectra on all major polymers are discussed by McCrum et al. [1]. Interpretations of the spectra are found in texts [2-4] and review articles [7,8].

The loss factors are most sensitive to the molecular motions. For example, in the glassy region the dynamic modulus shows small changes, yet the loss factors exhibit damping peaks.

The largest loss peak is generally the α-peak, which is associated with the glass transition temperature. This transition occurs in the amorphous regions of the polymer with the initiation of the micro-Brownian motion of the molecular chains. The magnitude of α-peak in the amorphous polymer is much higher than in the semicrystalline polymer, because the chain segments of the amorphous polymer are freer from restraints imposed by crystalline polymers in the glass transition region.

The glass transition temperature is affected by a number of chemical and molecular structures. Flexibility of the molecular chain, bulkiness of the side groups attached to the backbone, and molecular polarity affect the value of T_g, as discussed in Chap. I. The peak height and the temperature at which the α-peak occurs are closely related to the glass-transition phenomenon. Therefore, similar interpretations were used for both the α-peak and T_g.

The magnitude of the β-peak is small in comparison with the α-peak. The temperature at the β-peak is about 0.75 T_g. This dispersion is associated with an amorphous phase relaxation in most polymers. It is resolved into a double or triple peak and appears to involve local motion in the chain. In the case of amorphous polymers, the β-peak is very broad and may appear as a shoulder of the α-peak. It is associated with motion about the chain backbone of a relatively small number of monomer units or with motion of side groups [7].

The γ-peak temperature is below the β-peak temperature. In this region, the chain segments are frozen in, while side-group motion is made possible by defects in packing or configuration in the glassy and crystalline states. This γ-peak of polymers can be related to the following general classifications: side-group motions in the amorphous and crystalline phases, end-group rotation, crystalline defects, backbone-chain motions of short segments or groups, and phase separation of impurities or diluents [2,7].

The γ-relaxation in both amorphous and crystalline polymers could in many cases be attributed to a restricted motion of the main chain which required at least four $-CH_2$ groups in succession on a linear part of the chain [13]. This is the crankshaft mechanism of Shatzki [14] and Boyer [15]. It has been proposed that this mechanism is relevant to the γ-peak in polyethylene, polyamids, polyoxymethylene, and polypropylene oxide.

The α' (α_c) peak occurs at temperatures between T_g and the melting temperature. This process relates rotation about the chain axis of chain-folded polymers to a second-order expansion of lattice parameters perpendicular to the chain direction in the crystalline phase.

In the case of amorphous polymers the α' dispersion is often designated the $T_{\ell\ell}$ relaxation, and involves a premelting process. The $T_{\ell\ell}$ relaxation is exhibited in polystyrene [16,17], polyvinyl chloride [18], and polymethyl mechacrylate [19]. It has been extensively studied in synthetic rubber blends of the polybutadiene type [20]. In this study, it was found that $T_{\ell\ell}$ dispersion is related to the motion of the low-molecular weight components of a polymer with a broad molecular-weight distribution.

The relaxation spectra of nylon 6 and 66 are shown in Figs. 3.3 and 3.4, respectively. The results for tan δ on nylon 6 are for three

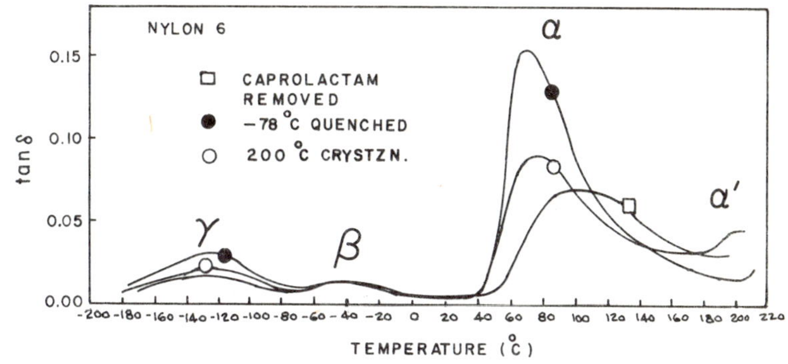

Fig. 3.3. Tan δ vs. temperature for three nylon 6 samples prepared by the different method, Data at 100 Hz. After Takayanagi [21].

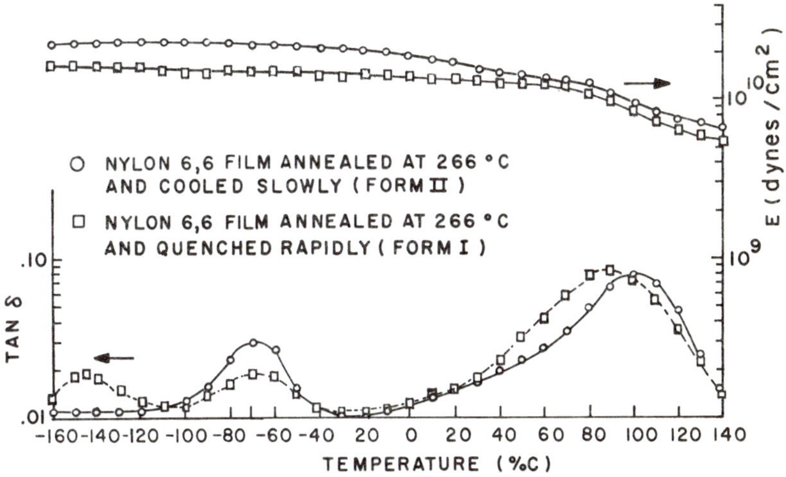

Fig. 3.4. E' and tan δ vs. temperature for nylon 66.

samples of different crystallinity [4]. The least crystalline specimen was obtained by quenching from the melt to -78°C, the sample of intermediate crystallinity was prepared by crystallizing at 200°C, and the preparation of the sample of highest crystallinity involved the removal of residual monomer (ε-caprolactam). The nylon-66 sample was prepared by quenching rapidly after annealing unoriented nylon 66 film at 266°C [21].

Three loss peaks are observed in Figs. 3.3 and 3.4. These have been labeled α, β, and γ and occur at approximately 90° to 110°C, -70°C to -40°C, and -120°C to -140°C, respectively. In addition to these three peaks, a further relaxation region in nylon 6 is indicated by the shoulder on the high-temperature side of the α-peak. Since this shoulder is most prominent for the specimen of highest crystallinity, it has been attributed by Takayanagi to a relaxation process involving the crystalline regions. He has also indicated that this relaxation might be related to a crystal-structure change from monoclinic to hexagonal, apparent from x-ray and specific-volume—temperature data. An additional crystalline relaxation mechanism might also be operative in the region of 190°C; the detection of this relaxation might be hindered by the relatively low crystalline content of the polyamides.

Figure 3.3 shows that the α-peak for nylon 6 decreases in height with increasing crystallinity, suggesting that the α-process involves motions within the amorphous phase. The α-damping-peak magnitude and temperature for nylon 6 and 66 have been shown to be sensitive to both the chemical composition of the nylon and the effect of moisture. The β-damping peak is atrributed to the carbonyl group of the polyamide forming hydrogen bonds. The temperature of this β-peak is influenced by the nylon moisture content [21]. Since nylons absorb some moisture from environmental humidity, moisture absorption characteristics must be considered in prospective uses of these materials. The low-temperature γ-transition is attributed to the crankshaft rotation of three or more methylene carbon atom chains between the amide groups.

Determination of the three relaxation spectra of polymers is useful not only for understanding the molecular motions in the structure, but also for practical applications. The α-dispersion is associated with the glass transition temperature and is accompanied by the greatest change in the stiffness of materials. Nearly all tough ductile glassy polymers and those with high impact strength have prominent secondary transitions (β- and γ-peaks) [2].

4. CRYSTALLINITY AND MOLECULAR ORIENTATION

Many polymers, including polyethylene terephthalate and nylon, crystallize if they are cooled from the melt. The polymer crystalline aggregates are known as spherulites. The chain-folding of molecules in lamellar crystals was discussed in Chap. I.

The degree of crystallinity and molecular orientation are affected by heat treatment and stress in the polymer processes. These structural changes have profound effects on the dynamic mechanical properties.

The dynamic modulus generally increases with increasing degree of crystallinity. This effect is clearly seen in the modulus above T_g. Nielsen has derived the following expressions for the relationships between the modulus above T_g and the degree of crystallinity W_c up to 70% [22]:

$$\log_{10} G = 6.763 + 4.77\, W_c, \tag{3.1}$$

$$\log G = W_c \log G_c + (1 - W_c) \log G_a, \tag{3.2}$$

where G is the dynamic shear modulus of semicrystalline polymers, G_c is the modulus of the pure crystal, and G_a is the modulus of the amorphous polymer above T_g. This equation implies that $G_c = 3.4 \times 10^{11}$ dyn/cm^2 for polymer crystals, and $G_a = 5.7 \times 10^6$ dyn/cm^2 for the rubbery phase. Equation (3.2) is called the logarithmic rule of mixtures, and it is known to hold for many types of two-phase systems in which both phases are continuous [2].

The crystallinity of many polymers has a great effect on the α-dispersion. The commonly used polymer polyethylene terephthalate (PET) is a good example of the changes of the α-peak due to crystallinity and molecular orientation. Controlling these parameters led to the introduction of a number of new products which possess a variety of mechanical properties. Some of the detailed discussions on PET are as follows:

(1) Effect of annealing temperature on crystallinity and crystal size of PET. The effect of structural changes on the height and position of the α-dispersion in PET has been extensively studied [23-27]. The dynamic mechanical properties of PET measured by the method of nonresonant forced vibration imply that the α-transition shifts from 80°C to 125°C with increasing crystallinity [28]. An investigation by a cantilever vibration method showed that although the α-transition initially shifted to higher temperature with increasing crystallinity, above a certain crystallinity the α-transition began to shift back [29]. This shifting back was also found by Takayanagi et al. [30], who used a Rheovibron dynamic viscoelastometer. In addition, they found a linear relation between the height of the α-dispersion peak and crystallinity. As the crystalline T_g increased, the height of the α-peak decreased.

The intensive study of the dynamic properties of unoriented films of different degrees of crystallinity by a torsion-pendulum method showed that the α-peak shifted to higher temperatures for crystallinities up to 30%. At higher crystallinities it shifted toward lower temperatures [31].

Illers and Breuer attributed this behavior to the effect of crystal size on the amorphous regions. At low to medium crystallinities there would be many small crystallites, which would act like crosslinks and inhibit the motion of segments in the amorphous regions, while at high crystallinities the crystallites would be larger and fewer, and therefore would allow the segments in the amorphous regions more freedom.

The degree of crystallinity is generally measured by x-ray diffraction techniques [32,33] or by measuring the density [34]. Morphological structure, including the length of chains between folds in crystals and spherulitic structure, can be measured by light scattering [35,36], small-angle x-ray scattering [37], and electron microscopy [38,39].

The most intensive study, a combination of x-ray diffraction and dynamic mechanical measurements for a series of unoriented PET fibers of different degrees of crystallinity, was made by Dumbleton and Murayama [40]. In this study, unoriented spun PET fiber of low crystallinity (≃2-3%) was used. Samples of different degrees of crystallinity were prepared by heating freely hanging skeins of yarn for 6 hours in an oven with a nitrogen atmosphere. Essentially all the crystallization occurred in the 6-hour period.

X-ray measurements of crystallinity were made by the procedure of Dumbleton and Bowles [41]. Wide-angle measurements of crystal size were made with an AEG focusing camera using crystal-monochromatized radiation. The resulting photographs were scanned by a Joyce-Loebel microdensitometer. Since the samples were unoriented, the meridional peaks were too weak to be detected and the equatorial peaks were overlapped by other diffraction maxima. Measurements were made on the (010) peak, as this reflection is only overlapped by the $(01\bar{1})$ peak. These two peaks were resolved graphically from the densitometer traces. The integral breadth, β, of the (010) peak was used to obtain the crystal size from the Scherrer formula [42]

$$D = \frac{K\lambda}{\beta \cos \theta}, \tag{3.3}$$

where λ is the wavelength of the x-rays, D is the crystal size, and θ is half the Bragg angle. The constant K, which depends on the method of measurement of the width of the diffraction peak and also on the shape of the crystals, was taken as unity.

Small-angle scattering measurements were made with pinhole collimation at a specimen-to-film distance of 32 cm. Ni-filtered Cu Kα radiation was used. The small-angle photographs were scanned with the microdensitometer, and from the resulting traces the position of the small-angle maximum was determined. The long period, d, was calculated from

$$d = \frac{\lambda}{2 \sin \theta}, \tag{3.4}$$

where θ is half the Bragg angle at which the small-angle maximum occurred.

The effect of annealing temperature on crystallinity for unoriented PET fibers is shown in Table 3.1. No crystallization was detected

TABLE 3.1.

X-ray crystallinity and vibron data for unoriented annealed PET fibers

Annealing temperature (°C)	Temperature at which maximum occurs in E'' (°C)	E''_{max} (10^9 dyn/cm^2)	Degree of crystallinity X (%)	Width of E'' peak at $E_{max}/\sqrt{2}$
85	83	2.5	0	6
97	84	2.2	0	8.5
98.5	87	2.1	1.5	6
100	90	1.75	11.5	8
101	95	1.62	18.5	14
102	97	1.50	25.5	20
110	98	1.40	31.0	21
125	100	1.35	35.3	25
155	100	1.25	42.2	26
176	100	1.23	49.0	26
200	98	1.25	49.7	26
228	95	1.20	48.0	20
245	92	1.23	48.2	21

until an annealing temperature of 98°C was reached, even though the glass transition of amorphous PET is 67°C [43]. The results show that in the temperature range 67°C to 90°C, parallelism of the benzene rings can occur without the formation of crystallites [44]. The degree of crystallinity of PET increases rapidly in temperature range from 98°C to 170°C. This behavior is a consequence of two crystallization processes: the formation of nuclei and their subsequent growth. The first process is dominant at low annealing temperatures, the second at high annealing temperatures. At intermediate temperatures both processes can occur. At low annealing temperatures crystallization will proceed via the formation of new crystallites rather than by the growth of existing crystallites. Thus, the crystal size should be almost independent of temperature in this region. At high temperatures the crystallization is by a diffusion process in which crystallites grow in size with little or no new nucleation.

The long period d obtained from small-angle x-ray scattering measurements is shown in Fig. 3.5 as a function of annealing

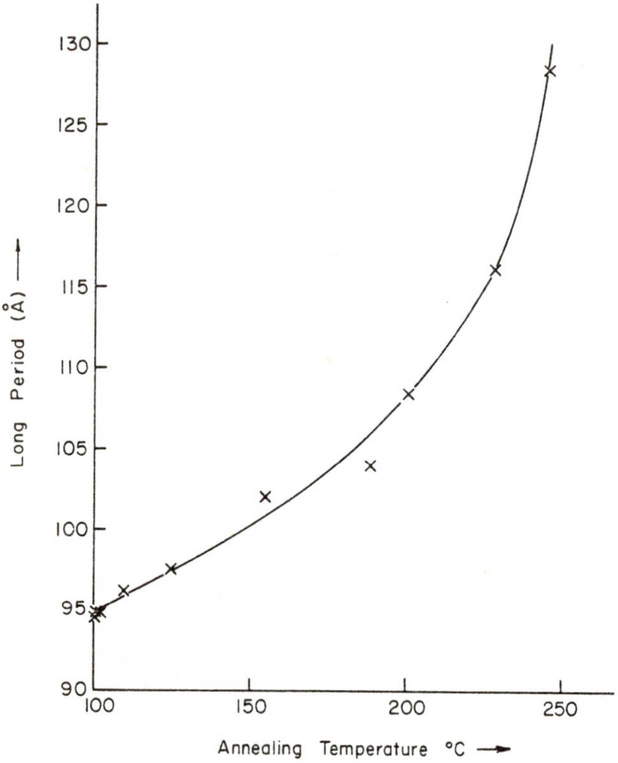

Fig. 3.5. Long period vs. annealing temperature.

temperature. At low annealing temperature it changes slowly with temperature, while for annealing temperatures near the melting point it increases rapidly. These results show that the increase in d is due to increase in crystal size.

Crystal sizes obtained from measurements on (010) are shown in Table 3.2. The crystal sizes of PET increase with increasing annealing temperature from 102°C to 245°C. These crystal sizes are weight-average sizes which employ integral breadths obtained by wide-angle x-ray measurements.

TABLE 3.2

Crystal-size data for unoriented PET fibers

Annealing temperature (°C)	Integral breadth (c)	Crystal size (Å)	Crystal volume (10^{-3} Å^3)	Amorphous volume/crystal (10^{-3} Å^3)	X/v (10^7 Å^{-3})	Long Period (Å)
85						-
97						-
98.5	a			17850	0.55	-
100	a			2100	4.22	94.5
101	a			1200	6.80	94.8
102	-0239	64.9	272	795	9.37	94.8
110	-0237	65.0	274	610	11.3	96.2
125	-0233	66.3	290	531	12.2	97.6
155	-02175	70.9	355	486	11.9	102.2
176	-02145	71.8	369	385	13.3	104.0
200	-0201	76.9	452	457	11.0	108.4
228	-0182	84.8	608	659	7.89	116.2
245	-0160	96.6	900	967	5.35	128.5

[a] Peak too weak to measure. Crystal size assumed equal to that of 102°C sample.

The true number-average size of crystals can be related to the weight-average size if one knows the correction for lattice distortion. A variety of measurements on polymers indicate that the lattice distortion should be approximately 1%. The volume of a crystallite as a function of annealing temperature in unoriented PET is shown in Table 3.2. This was calculated from the crystal size. It was assumed that the crystal volume is equal to the cube of the crystal size obtained from the (010) reflection. Ideally, crystal sizes should be obtained from several reflections so that the shape of the crystal may be obtained. In the absence of such information it is reasonable to take the shape of a crystal to be a cube.

(2) Relationships between crystallinity and dynamic mechanical properties. The dynamic mechanical properties of the same series of unoriented annealed PET fibers have been examined with a Rheovibron [4] which was operated at a frequency of 11 Hz. Samples were heated at 1°C/min in a nitrogen atmosphere under relaxed conditions, and measurements of the tensile modulus E and the loss factor tan δ were made at increments of 5°C except in the transition region, where smaller increments were used. Samples were allowed to equilibrate at temperature for 5 minutes before measurements were made. Measurements were continued up to the temperature at which the sample had been annealed. The annealing temperature was not exceeded, because this would have caused additional crystallization.

The temperature dependence of the real part of the modulus, E', for this series of PET is shown in Fig. 3.6. Changes in crystallinity

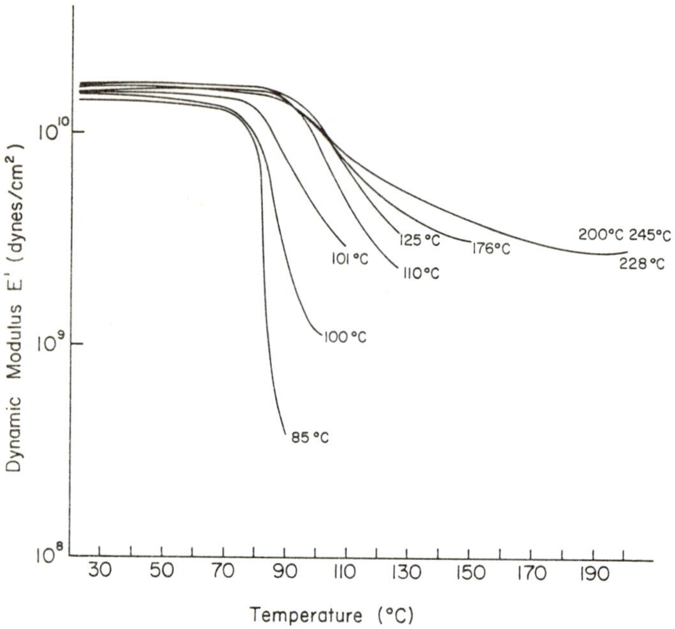

Fig. 3.6. Dynamic modulus E' as a function of temperature for PET of various crystallinities (11 Hz).

only slightly alter the room-temperature value of E', but have a great effect on the temperature dependence of E'.

The values of the maximum loss modulus, E''_{max}, and the temperature at which the peak occurs are given in Table 3.1. Also included are the half-power widths of the peak in the loss modulus. Figure 3.7 shows E''_{max} plotted against crystallinity. As crystallinity increased E''_{max} decreased. The width of the loss-modulus peak generaly increase as crystallinity increased, although at higher crystallinities it decreased somewhat. The increase in width is taken to represent the presence of an increased range of order. The temperature at which the maximum in E" occurs is plotted versus crystallinity in Fig. 3.8. The shift of E''_{max} to lower temperature at the higher crystallinities is evident.

In Fig. 3.9 a graph is shown of E" against temperature for the various unoriented annealed PET fibers, while in Fig. 3.10 a similar graph is shown of tan δ versus temperature. The intensity of tan δ

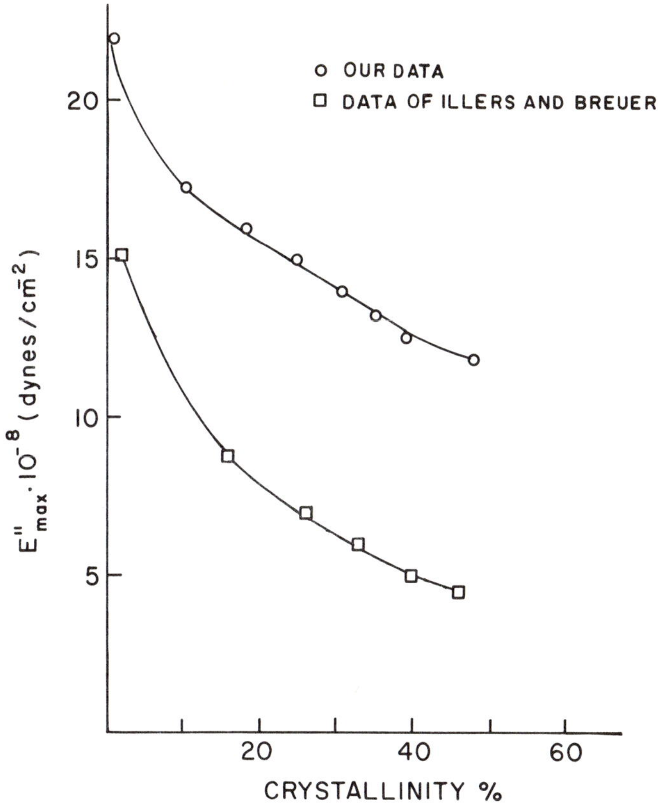

Fig. 3.7. The α-peak (E''_{max}) vs. crystallinity.

decreases with increasing degree of crystallinity. The intensity of the damping peak (tan δ) often is given by the equation [45]

$$\tan \delta = W_c (\tan \delta)_c + (1 - W_c) (\tan \delta)_a, \tag{3.5}$$

where W_c is the degree of crystallinity, and the subscripts a and c refer to the contributions of the amorphous and pure crystalline phases, respectively. Since the loss tangent (tan δ), in general, is mostly due to the amorphous phase, the tan δ equation for the maximum in the tan δ peak simplifies to

$$\tan \delta \simeq (1 - W_c) (\tan \delta)_a. \tag{3.6}$$

This equation shows the intensity decreases with increasing degree of crystallinity.

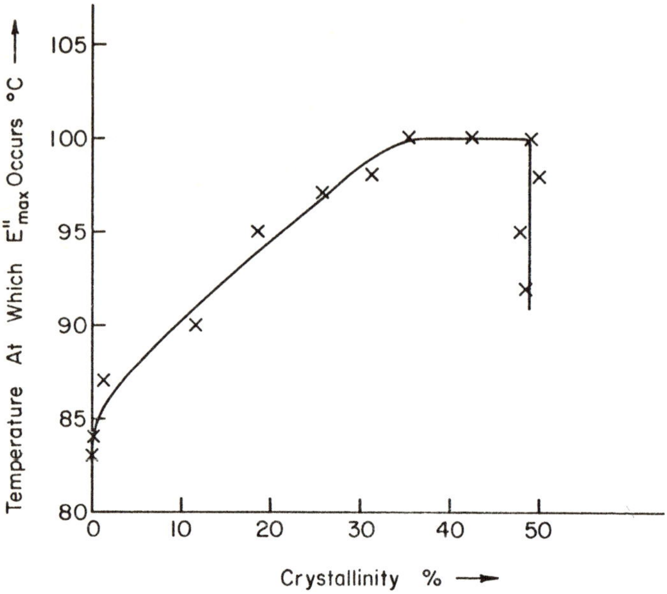

Fig. 3.8. Temperature at which E''_{max} occurs vs. crystallinity.

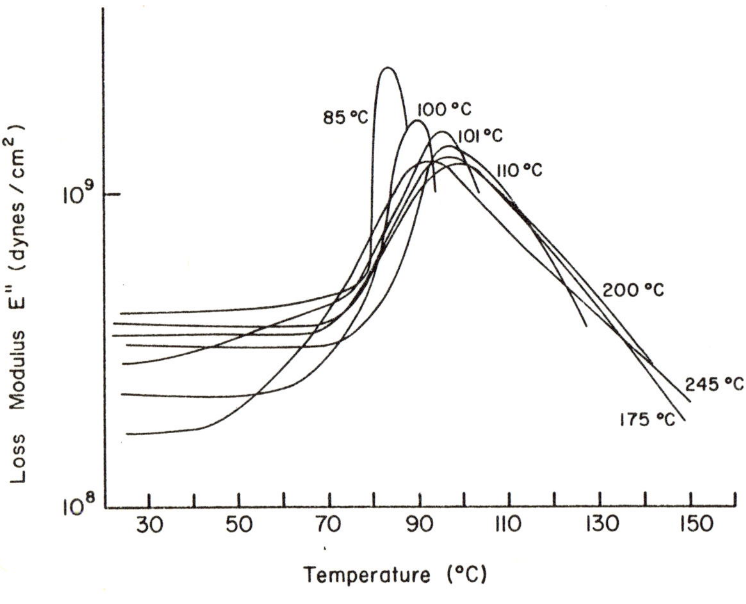

Fig. 3.9. Loss modulus E" as a function of temperature for PET of different crystallinities (11 Hz).

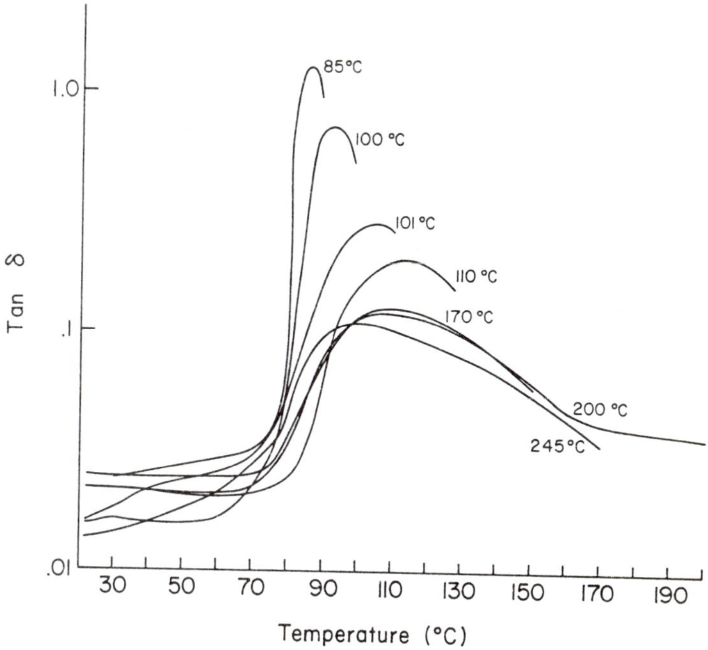

Fig. 3.10. tan δ vs. temperature for PET of different crystallinities (11 Hz).

(3) Interpretation of changes in position of the loss modulus peak (E''_{max}). As seen in Fig. 3.8, the temperature at which the maximum in the loss modulus E'' occurs is raised about 20°C by increasing the crystallinity from 2% to 50%, but at the higher crystallinity the peak temperature was lowered about 8°C. The shift in E''_{max} was a consequence of the inhibiting effect of crystalline regions on molecular motion in the amorphous regions. At constant crystallinity, many small crystallites are more effective in restraining motion than are a few large crystallites.

The amorphous volume associated with a crystallite will vary with crystallinity and crystal size. At low annealing temperatures, where the crystallinity increases rapidly with temperature, the amorphous volume per crystallite should decrease, while at high annealing temperatures, where the crystallinity remains constant, the amorphous volume per crystallite should increase. From simple considerations, an expression can be obtained for the amorphous volume per crystallit[e] [40]. The resulting values are given in Table 3.2. Figure 3.11 show[s]

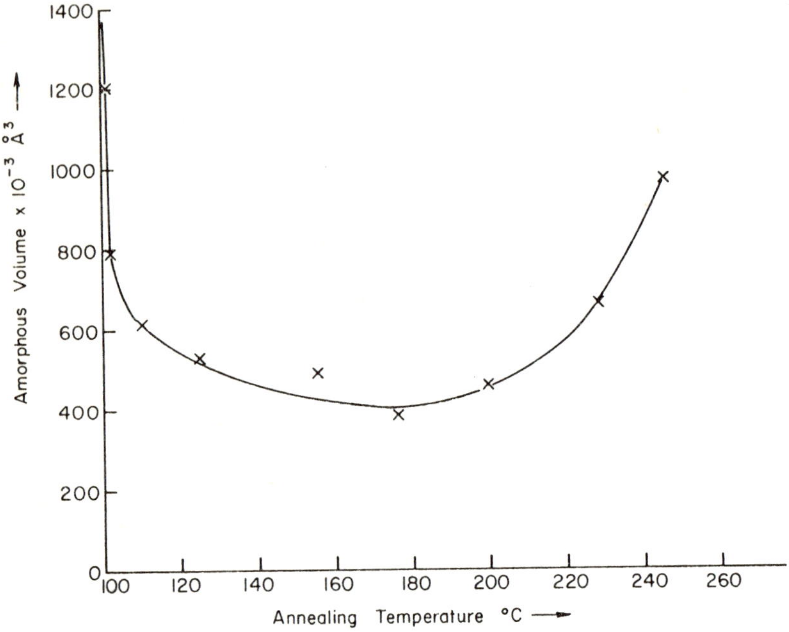

Fig. 3.11. Amorphous volume per crystallite vs. annealing temperatu[re]

the variation of amorphous volume with annealing temperature. It is seen that the amorphous volume first decreases as the annealing temperature increases, corresponding to the shift of E''_{max} to higher temperature, and then increases again, corresponding to the shift back in E''_{max}.

Roe, Smith, and Krigbaum [46] have postulated that the buildup of strain in the amorphous regions limits the degree of crystallinity which can be achieved at any given temperature. However, for the samples studied here the crystallinity is constant for annealing temperatures above 175°C, and the shift of E''_{max} to lower temperatur[e] for high crystallinities is indicative of less strain in the amorpho[us] regions. The limit of attainable crystallinity in this case must be due to the impingement of crystallites on one another and to the lac[k] of nuclei for growth at the higher annealing temperatures.

Expressions were developed by Uematsu and Uematsu [47] to quantitatively predict the change in position of E''_{max} with crystallinity.

Illers and Breuer [31] tried to apply this approach to their data without success. An alternative expression has been developed by Dumbleton and Murayama [40] on the assumption that the presence of a crystal reduces the free volume in the amorphous regions. It is found that the shift in E''_{max} is given by

$$\Delta T = \frac{\Lambda}{\alpha} \left(\frac{X}{v}\right) \frac{1}{P_c \bar{v}_o} \qquad (3.7)$$

where Λ is the decrease in free volume per crystallite, α is the thermal expansion coefficient of the free volume, X is the mass crystallinity, v is the mean volume of a crystallite, P_c is the density of the crystalline regions, and $\bar{v}_o$ is the occupied volume. The data are plotted in Fig. 3.12 as ΔT vs. X/v. There is good agreement

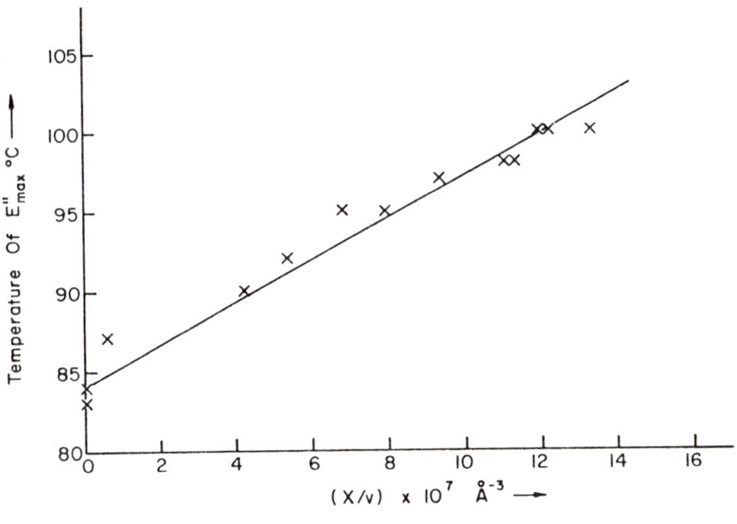

Fig. 3.12. Change in position of E''_{max} vs. X/v.

between theory and experiment, and the equation successfully predicts the shift of E''_{max} to lower temperatures.

It is interesting to note that polymers other than PET do not show great changes in the position of the loss-modulus peak with crystallinity. Isotactic polystyrene [48] and polypropylene [49,50] exhibit only minor changes of T_g with crystallinity. This is also shown by dynamic mechanical studies on poly(trifluorochlorethylene). With polyethylene and poly(tetrafluoroethylene) the glass transition temperatures are difficult to identify, and some polyamides, such as nylon 66, cannot be obtained with crystallinities varying over a wide range. On the basis of the present treatment, if ΔT is small, then either Λ is small or X/v remains constant; i.e., the number of crystals in the sample does not change appreciably with change in crystallinity.

(4) Effect of molecular orientation on the α-peak position. When a polymer is stretched, the molecules may be preferentially aligned along the stretch direction. In order to determine the effect of molecular orientation, a series of PET yarns were stretched with draw ratios from 2X to 5.5X, by drawing over a hot pin at 90°C. For these samples dynamic measurements using a Rheovibron were made at temperatures up to 160°C. The results are given in Table 3.3. The

TABLE 3.3
Vibron and x-ray data for drawn PET fibers

Draw Ratio	Position of maximum in E" (°C)	E''_{max} (10^9 dyn/cm^2)	Degree of crystallinity (%)	Width of E" peak
1	84	5.2	0	5
2	90	4.8	5.5	10
3	95	4.2	13.5	20
4	118	4.9	21.0	46
5	128	4.6	33.0	52
5.5	132	4.0	42.0	53

values of E" versus temperature are plotted in Fig. 3.13 [40]. The shift in E''_{max} for the drawn series is much greater than for heat-crystallized unoriented fibers. For example, the sample of 5.5X draw ratio has a crystallinity of 42%, which would indicate a change in position of E''_{max} from 80°C to around 100°C. However, the maximum occurs at 132°C. The results cannot be treated in the same way as for the undrawn series, since the drawn samples were not thermally conditioned. Some additional crystallization will occur on heating in the Rheovibron, as well as some loss of orientation as the samples are heated under relaxed conditions. These changes would tend to decrease the shift in E''_{max}.

Two effects may contribute to the greater shift of E''_{max} for the drawn samples. The crystals in a drawn sample may be smaller than these in an annealed sample of the same crystallinity. Also, in the drawing process the amorphous regions become oriented, and this will greatly reduce the free volume. It cannot be said unequivocally

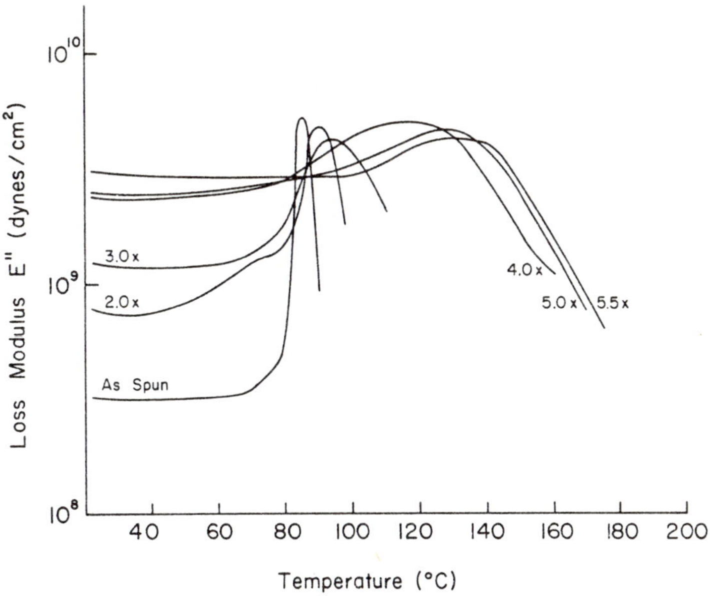

Fig. 3.13. Loss modulus E" as a function of temperature for PET of various draw ratios (11 Hz).

that the greater changes in the positions of E''_{max} are due to orientation alone. Newman and Cox [48] have shown that the combined effect of crystallinity and orientation is significantly greater for PET than for polystyrene.

The variation of dynamic modulus E' with temperature is given in Fig. 3.14. The modulus is an order of magnitude higher for the drawn series than for the heat-crystallized samples. There is also a greater change of E' within the drawn series.

The effects of the orientation in the crystalline and noncrystalline regions on the α-transition have been studied by using x-ray and birefringence on PET [51]. In this study PET yarns drawn 3X and 4.25X over a hot pin at 80°C were employed. Samples were heated for 6 hours at temperatures up to 240°C in vacuo under conditions in which the samples were free to shrink. All samples were subsequently boiled in water, in a relaxed condition, for 1 hour.

Dynamic measurements were made with a Rheovibron operated at a frequency of 11 Hz. Crystallinity measurements were made with a Norelco x-ray diffractometer, which was also used to obtain the crystallite orientation from scans on the ($\bar{1}05$) reflection. The orientation parameter employed was f_c given by [52]

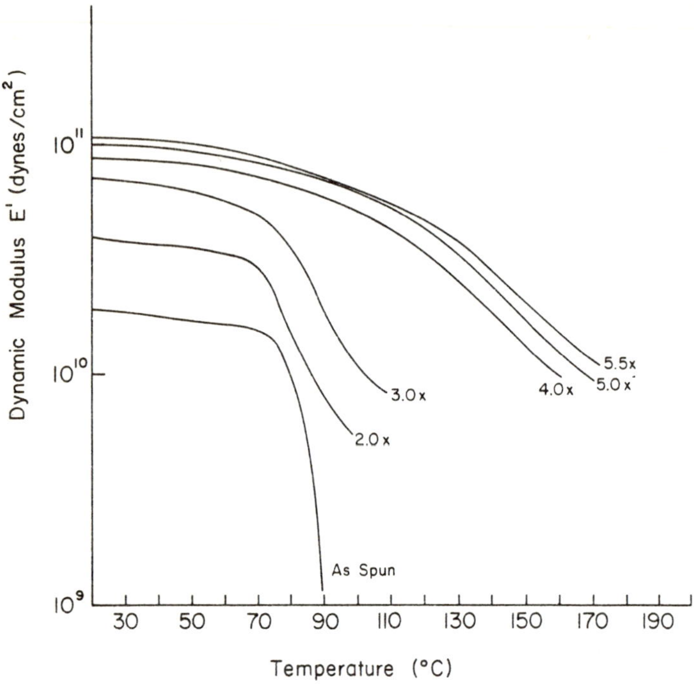

Fig. 3.14. Dynamic modulus E' as a function of temperature for PET of different draw ratios (11 Hz).

$$f_c = \frac{1}{2} (3[\cos^2 \theta] - 1), \qquad (3.8)$$

where $[\cos^2 \theta]$ is the mean-square cosine of the angle θ between a $(\bar{1}05)$-plane normal and the fiber axis. The orientation in the noncrystalline regions, f_{am}, was obtained from a combination of birefringence, crystallinity, and f_c [53]. Crystal-size measurements along the fiber direction were made from small-angle data collected with a Kratky camera using the method of Tsvankin [54].

X-ray and Rheovibron data for these drawn samples are given in Table 3.3. The crystallinity increases with annealing temperature, with higher crystallinities for the 4.25X than the 3X yarns, except at high annealing temperatures, where the crystallinities became similar. The orientation of the crystalline regions is almost the same for both draw ratios and is independent of annealing temperature. The orientation of the noncrystalline regions is also independent of annealing temperature, but is higher for the 4.25X yarns than the 3X yarns. The orientation of the noncrystalline regions is considerably lower than that of the crystalline regions. The long period in the small-angle scattering increases with annealing temperature and is higher for the 4.25X yarn series except at the highest annealing temperature. Crystal sizes calculated from the small-angle data increase with increased annealing temperature and are again somewhat higher for the 4.25X yarns except at the highest annealing temperature.

These results show that drawn crystalline PET yarns exhibit a shift in the position of the α-transition to lower temperatures for high annealing temperature; this effect has also been observed for unoriented crystalline PET. The main difference between the results for the drawn and unoriented yarns lies in the position of the α-transition on the temperature scale; the α-transition occurs at higher temperatures for higher draw ratios. The shift in the position of the α-transition due to annealing is interpreted in terms of the number of crystals, while the shift due to drawing is explained by orientation in the amorphous regions.

5. MOLECULAR WEIGHT AND CROSSLINKING

(1) Effect of molecular weight. Increasing molecular weight has a large effect in the glass-transition range, transforming the behavior from viscous flow to a plateau range of rubber-like behavior. The length of the rubbery plateau region increases as the molecular weight increases.

The loss tangent (tan δ) above T_g also is strongly dependent upon molecular weight. The value of the tan δ at the minimum above T_g decreases as the molecular weight increases [3,55,56].

The relationship between the dynamic mechanical properties and molecular weight for nylon 66 are discussed in a study carried out by Murayama and Silverman [57].

In order to prepare the high-molecular-weight nylon 66 film and fiber, thermally induced solid-state polycondensation (SSP) is applied. This procedure is described in detail by Silverman and Stewart [58,59]. Briefly, the process is carried out by heating polyamide fibers, such as polyhexamethylene adipamide fibers, with or without the usual additives (e.g., TiO_2, copper, potassium salts, phosphorus-containing compounds) to temperatures below their melting points (120-230°C) for a period of between 30 minutes and 24 hours in the absence of oxygen. Under these conditions, temperatures are high enough to allow the amine and acid ends of the polymer molecules to react, thus increasing the length of the molecules. Consequently, the molecular weight is increased. Time and temperature are dependent on each other inasmuch as higher temperatures require a shorter period of time to effect the desired result. It is most important, however, that the heating be carried out in the absence of oxygen and moisture. This may be done by any desired means—for example, a blanket of an inert gas, such as nitrogen, or a vacuum.

In accordance with these patents, films and oriented fibers composed of essentially linear polyamide molecules are produced. These materials are soluble in formic acid at 25°C. Table 3.4 shows the

TABLE 3.4
Conditions of thermally induced solid state polycondensation on nylon 66 and molecular weight

Sample		Conditions	Intrinsic viscosity η	Average molecular weight $\bar{M}_v$
Film	1	Control	1.6	45,000
	2	Extruded from resin, η = 2.0	2.0	60,000
	3	174°C, 24 hours, 8 mm Hg	3.5	120,000
Fiber	4	Control	1.31	35,000
	5	140°C, 18 hours, 8 mm Hg	1.6	45,000
	6	160°C, 8 hours, 8 mm Hg	1.7	48,500
	7	160°C, 18 hours, 8 mm Hg	1.81	52,000

conditions of the thermally induced solid-state polycondensation process for the test samples. The density did not change during treatment (1.14 g/cc); the fiber sample was wound on perforated metal bobbins with both ends of the threadline tied down to reduce relaxation.

The dynamic mechanical properties were measured at a constant frequency of 11 Hz with a Rheovibron instrument. Samples were heated at 1°C/min in a dry nitrogen atmosphere under relaxed conditions, and measurements of the tensile modulus E' and the loss tangent tan δ were made at increments of 10°C except near the transition region, when smaller increments were used. Samples were allowed to equilibrat at temperature for 5 minutes before measurements were made. The size of the film samples was about 0.5 cm X 5 cm X 0.1 cm, and the fiber samples were 800-1,000 denier.

Figures 3.15 and 3.16 show the loss tangent at 0 and 30% RH as a function of temperature for a control and for a typical thermally

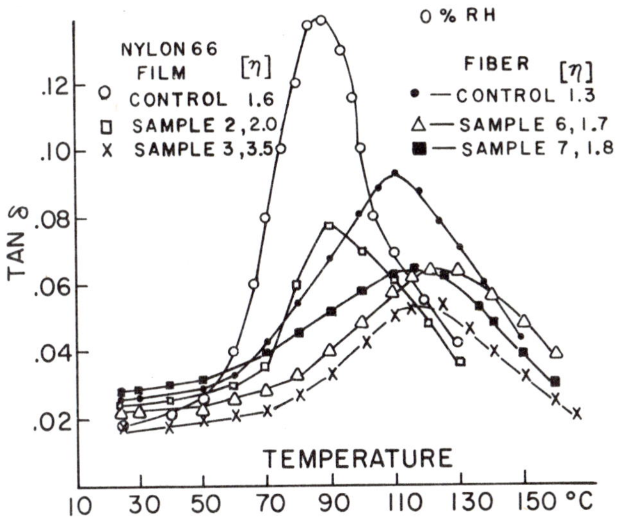

Fig. 3.15. Loss tangent of high molecular weight nylon 66 film and fiber as a function of temperature at 0% RH.

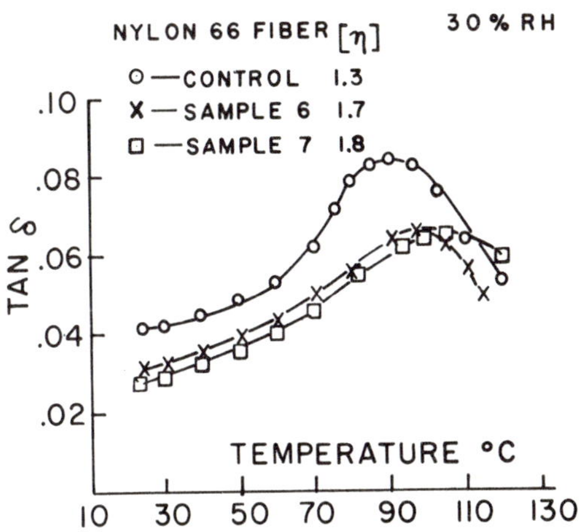

Fig. 3.16. Loss tangent of high molecular weight nylon 66 fiber as a function of temperature at 30% RH.

induced solid-state polycondensated nylon 66 film and fiber. The temperature of the maximum of the tan δ peak (T_g), the maximum tan δ value, and the width of the tan δ peak (ΔT) for these samples are summarized in Table 3.5. Figure 3.17 gives the relationship between

TABLE 3.5

Peak parameters of thermally induced solid-state polycondensation nylon 66

		0% RH			30% RH		
Sample		α-Peak temperature (°C)	$(\tan\delta)_{max}$	ΔT[a]	α-Peak temperature (°C)	$(\tan\delta)_{max}$	ΔT[a]
Film	1	88	.138	23	60	.104	57
	2	90	.080	40	72	.076	42
	3	120	.054	48	98	.082	62
Fiber	4	110	.094	34	90	.084	40
	5	118	.062	49	110	.062	46
	6	121	.063	64	103	.064	41
	7	123	.062	62	105	.065	50

[a]Width of α-peak at $(\tan\delta)_{max}/\sqrt{2}$.

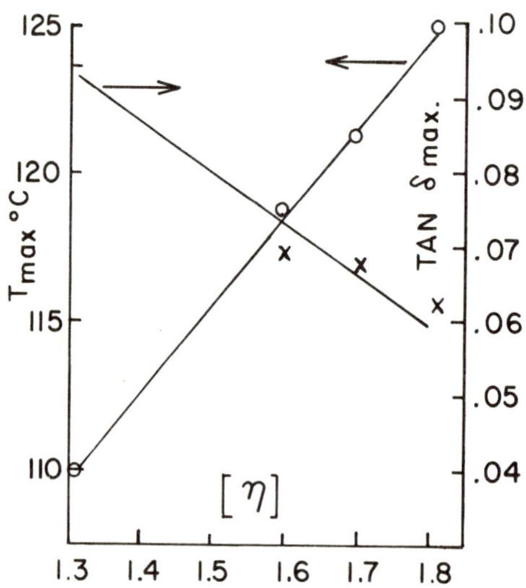

Fig. 3.17. Relation between intrinsic viscosity (η) and parameters of the α-dispersion (T_g and tan δ_{max}).

intrinsic viscosity (η) and the parameters of the α-dispersion [T_g and $(\tan\delta)_{max}$] in nylon 66 fiber. Intrinsic viscosities were measured in 90% formic acid at 20°C. The peak temperatures of tan δ for the treated films and fibers are shifted 8-32°C higher than those of the corresponding control nylon 66 (film and fiber), while the maximum height of the tan δ peaks is decreased. Figure 3.18 shows

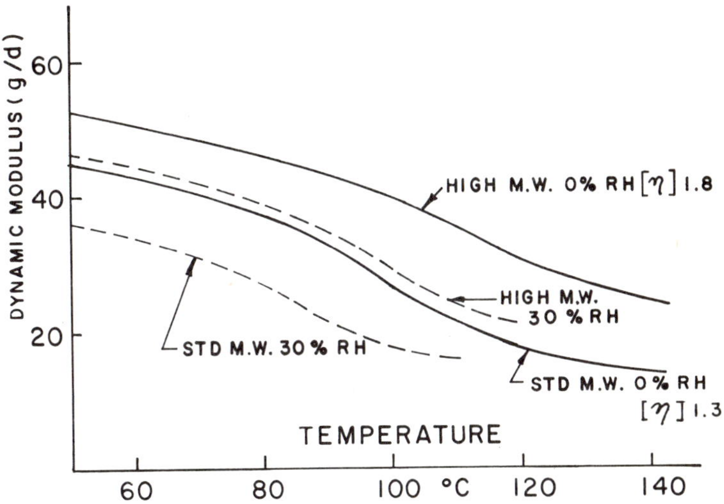

Fig. 3.18. Dynamic moduli of SSP fiber and control fiber at 0 and 30% RH.

the dynamic modulus (E') at 0 and 30% RH as a function of temperature for the high-molecular-weight fiber.

The occurrence of the α-transition, indicated by the maximum in tan δ at 110°C, for the highly oriented nylon 66 fiber is typical of the values obtained by other workers for similarly prepared nylon 66 [60-62]. The effect of annealing and drawing on the α-peak of nylon 66 is reported by Dumbleton and Murayama [62]. They conclude that the α-transition is little affected by the annealing treatment, but an increase in orientation causes the transition to move to a higher temperature. The results on dynamic mechanical properties of high-molecular-weight nylon 66 film and fiber clearly show that the α-transition is shifted to higher temperatures. The clearest and most dramatic effect of molecular weight on tan δ is shown in the unoriented films, whose intrinsic viscosities are 1.6, 2.0, and 3.5. In these curves, the loss of orientation occurring during SSP with its attendant effects in viscoelastic properties is not present to obscure the basic effects of molecular-weight increases. It should be noted that even the η = 3.5 film was free of crosslinking [58] and that the change in film properties is attributed to linear growth of molecules. The magnitude of the change in α-peak temperature as the intrinsic viscosity of the unoriented film is increased to 3.5 becomes apparent when one notes that this α-peak temperature becomes higher than that found for oriented nylon of normal molecular weight. A shift of the α-peak to high temperatures is usually a symptom of increasing crystallinity or more closely packed morphology. However, the study on the dynamic mechanical behavior of nylon 66 by Dumbleton and Murayama indicated that crystallinity is little affected by temperature from 100°C to 180°C (46.4-55% crystallinity). The α-peak in high-molecular-weight nylon 66 may lead to the formation of the additional interlamellar connections [63].

At 30% RH the peak temperature of tan δ for the nylon 66 control is approximately 20°C lower than it is at 0% RH. A similar shift is found by Woodward et al. [60]. They report the shift for nylon to be 15°C per percent of moisture. At 30% RH, the peak temperature of tan δ for the high-molecular-weight nylon 66 fiber is also shifted to lower temperature; however, the decrease in peak amplitude is less than that of the nylon 66 control. These phenomena suggest that the high-molecular-weight nylon 66 has less sensitivity to moisture.

(2) Effect of crosslinking. The effect of crosslinking on the dynamic properties of the vulcanized rubber is evident above T_g [2]. The dynamic modulus of a crosslinked rubber does not decrease with increasing temperature. The loss tangent is a sensitive indicator of crosslinking. At temperatures well above T_g, the damping decreases with increasing degree of crosslinking. Epoxies and highly crosslinked phenol-formaldehyde resins are examples of polymers which are much more highly crosslinked than normal rubbers.

Figures 3.19 and 3.20 are typical of how the dynamic mechanical properties change with high degrees of crosslinking [64,65]. The dynamic modulus above T_g dramatically increases while the α-peak (T_g) shifts to higher temperature with increasing crosslinking. These epoxy samples were Epon 828 reacted with various amounts of methylene diamine (MDA); the amount of amine was -6% to + 102% of the stoichiometric amount. The amine was melted and mixed with the diepoxy at 100°C. A sheet was then cast on a glass plate. After curing in an air oven 45 minutes at 60°C, then 30 minutes at 80°C, and finally $2\frac{1}{2}$ hours at 150°C, the samples were allowed to cool slowly to room temperature.

A plot of tan δ as a function of temperature is shown in Fig. 3.19. The effect of a greater than stoichiometric amount of amine is to decrease the temperature and increase the height of the loss peak.

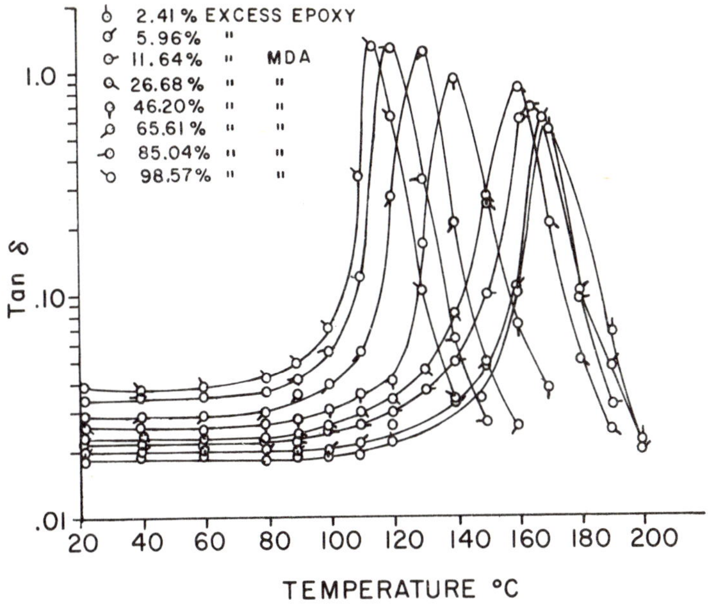

Fig. 3.19. Loss tangent as a function of temperature.

An excess of amine caused a significant shift in T_g, whereas a moderate excess of epoxy did not. This is reasonable when one considers that excess amine increases the molecular weight between crosslinks, M_c [69], whereas the M_c for all epoxy-excess polymers should be about the same; the polymers will differ only in regard to the number of large crosslinked molecules in the system and the number of dangling epoxy ends.

The elastic component E' of the complex modulus for these epoxy samples is shown as a function of temperature in Fig. 3.20. Again one observes the shift in the transition temperature as the stoichiometric excess of amine is increased, and also that the order of the transition temperatures is the same as the order of the moduli at room temperature; the highest glass-transition temperature is associated with the highest room-temperature modulus. The curves have the shape characteristic of crosslinked amorphous polymers in the transition region.

Crosslinking theory [66] requires a high degree of crosslinking to result in a higher value of E' after the transition than a low degree of crosslinking. In the present case we would expect that samples near 0 stoichiometric excess amine would have a higher degree

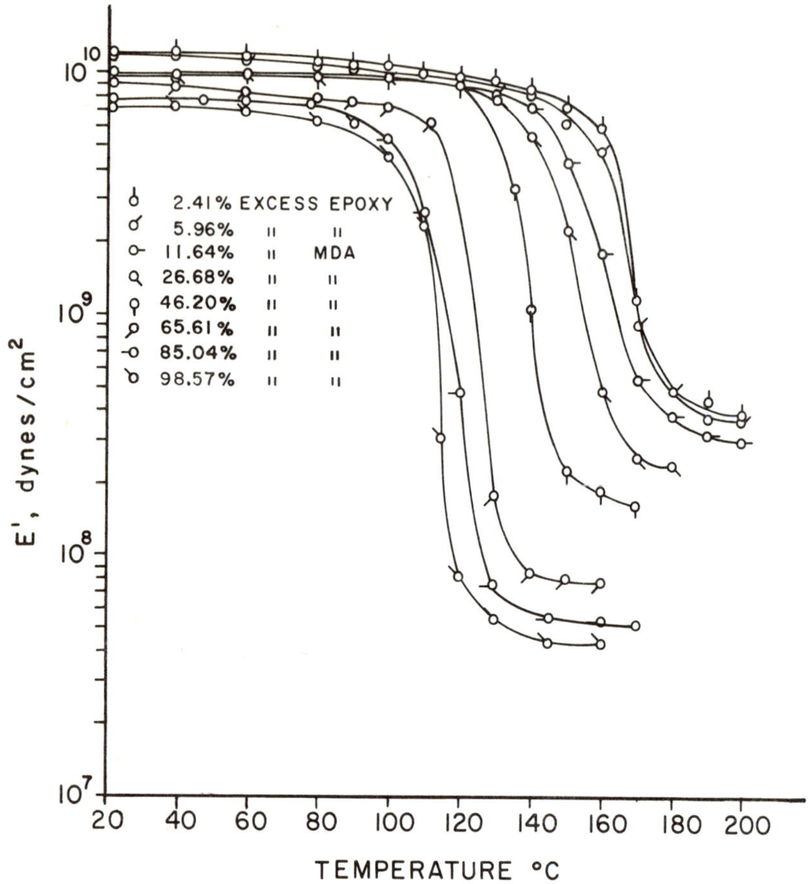

Fig. 3.20. Dynamic modulus E' as a function of temperature.

of crosslinking than the samples near 98.56%, and that they therefore would have a higher value of E' at $T_g + 40°C$. This is observed in Fig. 3.20.

The shear modulus G of crosslinked rubbers is given by [67]

$$G = \frac{\bar{r}_i^{\,2}}{\bar{r}_f^{\,2}} \frac{dRT}{M_c} \left(1 - \frac{2M_c}{\bar{M}_n}\right). \quad (3.9)$$

Here $\bar{r}_i^{\,2}/\bar{r}_f^{\,2}$ is the ratio of the mean square end-to-end distance of the polymer chain in the sample to the same quantity in a randomly coiled chain [67]. For the moment this ratio will be assumed to be equal to 1.0. d is the polymer density, R is the gas constant, T is the absolute temperature, M_c is the molecular weight between crosslinks and $\bar{M}_n$ is the chain-backbone molecular weight. The factor 1-2 X $M_c/\bar{M}_n$ is a correction for chain ends and becomes negligible when $\bar{M}_n$ is very large relative to M_c, as it is in the present case. Thus we have

$$G = \frac{dRT}{M_c}. \tag{3.10}$$

G must be measured in the rubbery plateau region above T_g. For present purposes, E' was measured at $T_g + 40°$. The shear modulus G can be taken as E'/3 [68].

The values for M_c in equation (3.10) have been obtained by Murayama and Bell [64]. The resulting values of G agree reasonably well with the reaction-stoichiometry values at low values of M_c, but are too high for samples with stoichiometric excess amine greater than 60%.

A number of studies have been made which relate the distance between crosslinks to the shift in T_g [70-74]. A rough estimate of M_c can be obtained from

$$M_c = \frac{3.9 \times 10^4}{T_g - T_{go}} \tag{3.11}$$

where T_{go} is the transition temperature of uncrosslinked polymer. To estimate T_{go}, a polymer was made by partially reacting the MDA and Epon 828 so that almost all of the primary amine groups, but very few of the secondary amine groups, were consumed; the result approximated the linear polymer. The transition temperature was estimated by heating rapidly and noting the apparent softening temperature under stress with a Vibron. The transition was found in the region 75-80°C. The M_c values calculated from the T_g shift are consistent with the stoichiometry [64].

6. POLYMER BLEND AND COPOLYMERS

The dynamic mechanical properties of polymer blends are determined primarily by the mutual solubility of the two homopolymers. If two polymers are compatible and soluble in one another, the properties of the blend are nearly the same as those of a random copolymer of the same composition. However, many mixtures of polymers form two phases, due to insolubility of polymers. In this case, the damping-temperature curve shows two peaks; each peak is characteristic of the glass-transition temperature of one of the components. The two steps in the dynamic-modulus—temperature curve are characteristic of an immiscible two-phase system [2]. Figure 3.21 shows the dynamic mechanical properties of a polyblend. In this sample which is called

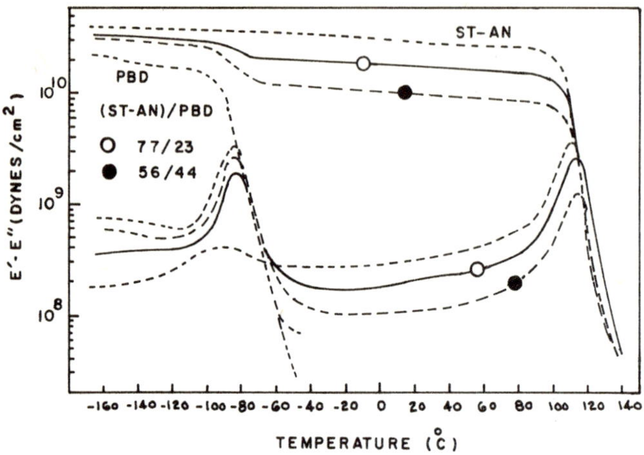

Fig. 3.21. E' and E" at 138 Hz vs. temperature for polymer of polybutadiene (PBD) with sytrene (ST) and acrylonitrile (AN) monomers. After Takayanagi [4].

ABS resin [75], polybutadiene (PBD) particles are dispersed in a phas of styrene-acrylonitrile copolymer (ST-AN, or SAN) with the formation of graft bonds on its boundary. The temperature dependence of the dynamic modulus E' and loss modulus E" of this polyblend is shown in Fig. 3.21. There are two blend ratios: 77/23 (white circles) and 56/44 (black circles), expressed as volume fractions of ST-AN/PBD. The results on the dynamic properties of PBD film and ST-AN copolymer at 138 Hz are also plotted as broken lines in Fig. 3.21 [4].

Two transition peaks are observed. The dispersion at 115°C is the α-peak of ST-AN copolymer, while the α-peak of PBD is at -82°C. These two polymer blends in fact exhibit dispersions at 115°C and -82°C. However, the intensity of the two peaks decreases with decreasing ST-AN or PBD in the polymer blends. The magnitude of each peak is a characteristic of the relative concentrations of the two components. This behavior is found whether the phases are dispersed or continuous [76-78].

Using these results, Takayanagi has calculated the theoretical moduli of an ST-AN/PBD blend from the equation of the mechanical equivalent model of two-phase mixtures [4,79]. This equivalent model is represented in Fig. 3.22. The left side of this figure is a schematic representation of the two-phase system, in which the particles (as phase 2) are dispersed in the medium (as phase 1).

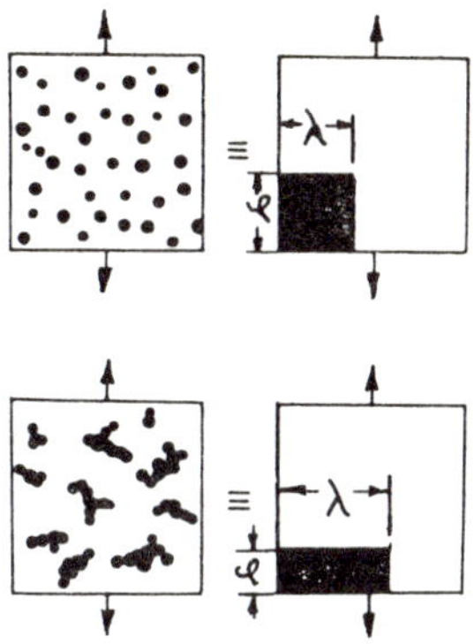

Fig. 3.22. Schematic diagram representing the relationship between the mixing state in polymer blend and λ and φ. After Takayanagi [4].

This system may be symbolized in the simplest case as shown in the right-hand side. λ and φ in the figure are parameters related to the mixing state and the composition of the sample. The mechanical models for calculating the modulus of the equivalent model are shown in Fig. 3.23. In model (a) in Fig. 3.23, the strain on an element

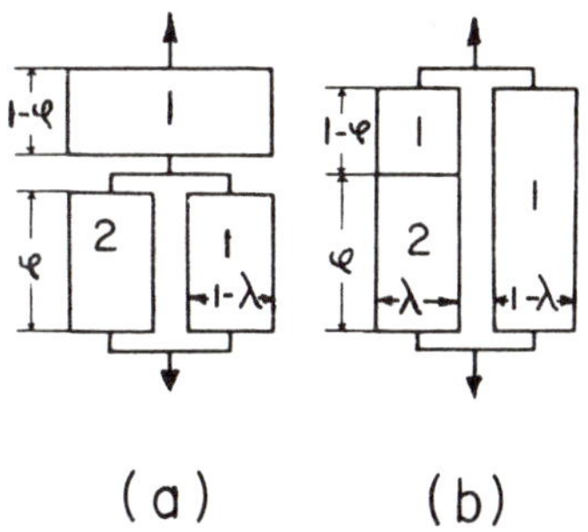

Fig. 3.23. Mechanical models for calculating modulus of equivalent model. After Takayanagi [4].

of phase 2 is the same as that on the right element neighboring it. In model (b), the strain on phase 2 is different from those on the other elements. It has been proved by model experiments that model (a) is more suitable for calculating the modulus in the equivalent model [4]. The complex modulus E* of this equivalent model is expressed by

$$E^* = \left[\frac{\phi}{(1 - \lambda) E_1^* + \lambda E_2^*} + \frac{1 - \phi}{E_2^*} \right]^{-1}, \qquad (3.12)$$

where E_1^* and E_2^* are the complex moduli of the medium and the dispersed particles respectively. λ and ϕ are parameters representing the mixing state, and they are so related that $\lambda\phi$ is equal to the volume fraction of the particles.

In the case of ST-AN/PBD blends, these parameters in equation (3.12) are as follows: E_1^* is the value for PBD; E_2^* is the value for ST-AN copolymer. For the volume fraction of PBD, $vR = \lambda\phi = 0.23$, $\lambda = 0.600$, $\phi = 0.380$. For ST-AN, $vR = 0.44$, $\lambda = 0.775$, $\phi = 0.568$. The calculated moduli are shown by the full lines in Fig. 3.21. Satisfactory agreement between the observed and the calculated curves is seen, except for a slight deviation in the dispersion regio

The dynamic mechanical properties of copolymers are strongly influenced by the characteristics of the homopolymers and the composition of the copolymer. In a homogeneous copolymer all the molecule have the same chemical composition, and hence should have a single sharp α-transition (T_g). Many copolymers are heterogeneous, that is, different molecules have different chemical composition. This heterogeneity of chemical composition comes about because one monomer tends to be more reactive than the other. At the beginning of polyme zation, one comonomer tends to concentrate in the copolymer more than the other. During the course of the polymerization the more reactive monomer is depleted in concentration, so the copolymer molecules become enriched with the slower-reacting monomer late in the reaction. Thus, the heterogeneous copolymer ends up with molecules which have a wide spread in T_g [2]. For this reason the damping curve of the dynamic properties of the heterogeneous copolymer reflects the broadening of the transition region due to the increase in heterogeneity [80]. The broadening of the α-peak is related to the mutual solubility of the two homopolymers. If the homopolymers tend to be insoluble in one another, then heterogeneous copolymers have broad transitions. If the molecular interaction between the two types of polymers is high, the effect of interchain chemical heterogeneity is small, so there will be a sharp peak in the transition region [2].

The relative concentration of the two copolymers or polyblends has a dramatic effect on the primary dispersion [4,81]. Figure 3.24 shows how the dynamic mechanical properties change typically with amounts of components. The samples of different compositions were prepared by vaporization of the tetrahydrofuran solution of polyviny chloride (PVC) and butadieneacrylonitrile copolymer rubber (NBR). The dynamic modulus and loss modulus are shifted on the temperature

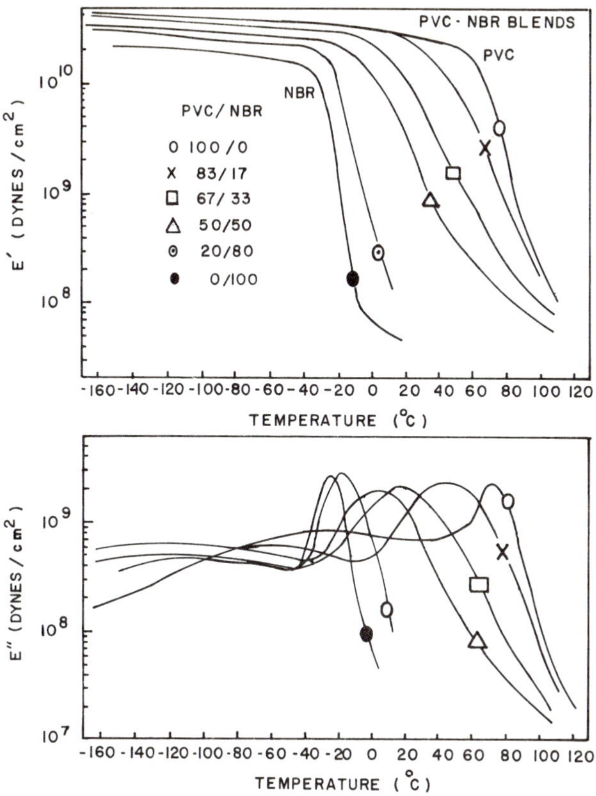

Fig. 3.24. E' and E" vs. temperature for blended system of PVC and NBR of various compositions at 138 Hz. After Takayanagi [4].

scale in proportion to the relative concentration of the two component polymers. The damping peaks in the intermediate compositions are broader in the transition region. This suggests the existence of microheterogeneity due to the different segmental environments. In the case of homogeneous copolymers the shape of the dispersion does not change, and only the position of the dispersion shifts with composition [82]. Similar effects on dynamic properties are observed with plasticizers [83].

Copolymerization and polymer blends are not only useful for modifying the properties of polymers, but also for improving the processability of the molten polymers.

REFERENCES

1 N. G. McCrum, B. E. Read, and G. Williams, Anelastic and Dielectric Effect in Polymeric Solids, Wiley, New York, 1967.
2 L. E. Nielsen, Mechanical Properties of Polymers and Composites, Marcel Dekker, New York, 1974.
3 J. D. Ferry, Viscoelastic Properties of Polymers, 2nd edn., Wiley, New York, 1969.
4 M. Takayanagi, Viscoelastic Properties of Crystalline Polymers, Mem. of the Fac. of Eng., Vol. 23, No. 1, 1963.
5 M. Takayanagi, in Proceedings of the Fourth International Congress of Rheology, Part I, Interscience, New York, 1965, p. 161.
6 M. Takayanagi and T. Matsuo, J. Macromol. Sci.--Phys., B1 (3) (1967) 407.
7 R. F. Boyer, in Polymeric Materials: Relationships between Structure and Mechanical Behavior (E. Bear and S. V. Radcliffe, Eds.), Am. Soc. Metals, Ohio, 1974, Chap. 6, pp. 227-368.
8 R. F. Boyer, Polymer, 17 (1976) 996.
9 K. Deutsch, E. A. W. Hoff, and W. Reddish, J. Polym. Sci., 13 (1954) 365.
10 J. G. Powles, B. I. Hunt, and D. J. H. Sandiford, Polymer, 5 (1964) 505.
11 K. M. Sinnot, J. Polym. Sci., 42 (1960) 3.
12 J. Heijboer, Physics of Non-Crystalline Solids, North-Holland, Amsterdam, 1965, p. 231.
13 A. H. Willbourn, Trans. Faraday Soc., 54 (1958) 717.
14 T. F. Shatzki, J. Polym. Sci., 57 (1962) 496.
15 R. F. Boyer, Rubber Rev., 34 (1963) 1303.
16 R. F. Boyer, Rubber Chem. Technol., 36 (1963) 1303.
17 R. F. Boyer, J. Polym. Sci., C. 14 (1966) 267.
18 L. Utracki, J. Macromol. Sci. Phys., B10 (3) (1974) 477.
19 K. Ueberreiter and J. Naghazadeh, Kolloid-Z. Z.-Polym., 250 (1972) 927.
20 E. A. Sidorovitch, A. I. Marei, and N. S. Gashtol'd, Rubber Chem. Technol., 44 (1971) 166.
21 J. P. Bell and T. Murayama, J. Polym. Sci., Part A-2, 7 (1969) 1059.
22 L. E. Nielsen, J. Appl. Polym. Sci., 2 (1959) 351.
23 G. Farrow, J. McIntosh, and I. M. Ward, Makromol. Chem. 38 (1960) 147.
24 D. E. Kline and J. A. Sauer, Polymer, 2 (1961) 401.
25 R. Meridith and Hsu Bay-Sung, J. Polym. Sci., 61 (1962) 271.
26 P. R. Pinnock and I. M. Ward, Proc. Phys. Soc., 81 (1963) 260.
27 P. R. Pinnock and I. M. Ward, Polymer, 7 (1966) 255.
28 A. B. Thompson and D. W. Woods, Trans. Faraday Soc., 52 (1956) 1383.
29 T. Kawaguchi, J. Polymer Sci., 32 (1958) 417.
30 M. Takayanagi, M. Yoshino, and S. Minami, J. Polym. Sci., 61 (1962) 171.
31 K. H. Illers and H. Breuer, J. Colloid Sci., 18 (1963) 1.
32 R. L. Miller, in Encyclopedia of Polymer Science and Technology, Wiley, New York, 1966, Vol. 4, p. 449.
33 R. L. Miller, Crystalline Olefin Polymers, Part 1 (R. A. V. Raff and K. W. Doak, Eds.), Interscience, New York, 1965, p. 577.
34 F. P. Price, J. Chem. Phys., 19 (1951) 973.
35 M. B. Rhodes and R. S. Stein, J. Appl. Phys., 32 (1961) 2344.
36 R. S. Stein, Polym. Eng. Sci., 9 (1969) 320.
37 G. Porod., Adv. Polym. Sci., 2 (3) (1961) 363.
38 P. H. Geil, Polymer Single Crystals, Interscience, New York, 1963.
39 A. Keller, Philos. Mag., 2 (1957) 1171.

40 J. H. Dumbleton and T. Murayama, Kolloid-Z. Z.-Polym., 220 (1967) 41.
41 J. H. Dumbleton and B. B. Bowles, J. Polym. Sci., 4 (1966) 951.
42 H. P. Klug and L. E. Alexander, X-ray Diffraction Procedures, Wiley, New York, 1954.
43 H. J. Kolb and E. F. Izard, J. Appl. Phys., 20 (1949) 564.
44 H. G. Kiliam, H. Halboth, and E. Jenckel, Kolloid-Z. Z.-Polym., 172 (1960) 166.
45 R. W. Gray and N. G. McCrum, J. Polym. Sci., A2, 7 (1969) 1329.
46 R. J. Roe, K. J. Smith, and W. R. Krigbaum, J. Chem. Phys., 35 (1961) 1306.
47 Y. Uematsu and I. Uematsu, Rep. Progr. Polym. Phys. (Jap.), 3 (1959) 27, 4 (1960) 102.
48 S. Newman and W. P. Cox, J. Polym. Sci., 46 (1960) 29.
49 J. Van Schooten, H. Van Hoorn, and J. Boerma, Polymer, 2 (1961) 161.
50 E. Passaglia and G. M. Martin, J. Res. Natl., Bur. Stand., 68A (1964) 519.
51 J. H. Dumbleton, T. Murayama, and J. P. Bell, Kolloid-Z. Z.-Polym., 228 (1968) 54.
52 R. J. Samuels, in Science and Technology of Polymer Films (O. J. Sweeting, Ed.), Wiley-Interscience, New York, 1968, Chap. 7.
53 J. H. Dumbleton, J. Polym. Sci., A-7, 7 (1969) 667.
54 D. Tsvankim, Ya. Polym. Sci. USSR, 6 (1964) 2304.
55 W. P. Cox, R. A. Isaksen, and E. H. Merz, J. Polym. Sci., 44 (1960) 149.
56 G. Pezzin, G. Ajroldi, and C. Garbuglio, Rheologica Acta, 8 (1969) 304.
57 T. Murayama and B. Silverman, J. Polym. Sci. Polym. Phys. Ed., 11 (1973) 1873.
58 B. Silverman and L. E. Stewart, U.S. Patent 3,562,206, 1971.
59 B. Silverman and L. E. Stewart, U.S. Patent 3,548,584, 1970.
60 A. E. Woodward, J. M. Crissman, and J. A. Sauer, J. Polym. Sci., 44 (1960) 23.
61 J. P. Bell and T. Murayama, J. Polym. Sci., Part A-2, 7 (1969) 1059.
62 J. H. Dumbleton and T. Murayama, Kolloid-Z. Z.-Polym., 238 (1970) 410.
63 A. Miyagi and Wunderlich, Bull. Am. Phys. Soc., Ser. 11, 16 (3) (1970) 410.
64 T. Murayama and J. P. Bell, J. Polym. Sci., A2, 8 (1970) 437.
65 M. F. Drumm, C. W. H. Dodge, and L. E. Nielsen, Ind. Eng. Chem., 48 (1956) 76.
66 P. J. Flory, Principles of Polymer Chemistry, Cornell U. P., Ithaca, N.Y., 1953, Chap. 11.
67 A. V. Tobolsky, D. W. Carlson, and N. Indictor, J. Polym. Sci., 54 (1961) 175.
68 A. V. Tobolsky, J. Polym. Sci., C, 9 (1965) 174.
69 J. P. Bell, J. Polym. Sci., A-2, 8 (1970) 417.
70 K. Ueberreiter and G. Kanig, J. Chem. Phys., 18 (1950) 399.
71 T. G Fox and S. Loshaek, J. Polym. Sci., 15 (1955) 371,391.
72 G. M. Martin and L. Mandelkern, J. Res. Nat. Bur. Stand., 62 (1959) 141.
73 H. D. Heinze, K. Schmieder, G. Schnell, and K. A. Wolf, Kautschuk Gummi, 7 (1961) 208; Rubber Chem. Technol., 35 (1962) 776.
74 E. A. DiMarzio, J. Res. Nat. Bur. Stand., 68A (1964) 611.
75 N. E. Davenport, I. W. Hubbard, and M. R. Putti, Br. Plast., Dec. 1959, p. 549.
76 T. Miyamoto, K. Kodama, and K. Shibayama, J. Polym. Sci., A2, 9(1971) 2131.

77 R. Buchdahl and L. E. Nielsen, J. Appl. Phys., 21 (1950) 482.
78 L. E. Nielsen, Appl. Polym. Symp., No. 12 (1969) p. 249.
79 M. Takayanagi, S. Uemura, and S. Minami, J. Polym. Sci., Part C, 5 (1964) 113.
80 L. E. Nielsen, J. Am. Chem. Soc., 75 (1953) 1435.
81 D. I. Livingston and J. E. Brown, Jr., in Proc. 5th Int. Congr. Rheol., Vol. 4 (S. Onogi, Ed.), Univ. Tokyo Press, Tokyo, 1970, p. 25.
82 A. V. Tobolsky, Properties and Structures of Polymers, Wiley, New York, 1960, p. 79.
83 K. Schmieder and K. Wolf, Kolloid-Z. Z.-Polym., 127 (1952) 65.

CHAPTER IV

ANALYSIS OF STRUCTURE GEOMETRY

1. INTRODUCTION

The effects of molecular parameters on the dynamic mechanical properties of polymers have been discussed in Chap. III. These variables are based on the internal structure of the polymers. However, many polymeric materials are combined as structural units. Examples are the laminated plastics, the fiber-reinforced composites, tires, and fabrics. Dynamic mechanical analysis is useful in determining the properties of these structural units and effects of the structural geometry on the dynamic properties. The end-use performance of these structural assemblies is influenced by the inherent properties of the components and the external structural factors of the systems.

The damping characteristics of structural systems have been the subject of investigation for many years. In the past several decades, engineers have expressed considerable interest in the damping (internal friction) properties of structures at stress levels encountered in engineering design. They have contended that resonant structural vibrations could be controlled by the effective use of this property.

In addition to the internal damping properties of structural materials, attention has been paid to the use of structural fabrications that involved Coulomb-slip and viscoelastic-shear damping mechanisms. With the advent of high-speed space vehicles, there was great need for degrees of damping in structures that exceeded what was available from the structural material itself. Investigators thus turned to such structural damping techniques as interfacial Coulomb-slip damping, viscoelastic laminate damping, and Coulomb or viscoelastic junction damping. Detailed studies of structural damping can be found in books from the ASME [1] and ASTM [2]. In this chapter the effect of the structural geometry on dynamic mechanical properties will be discussed in examples of interfaces in assemblies, adhesion, and composite structure.

2. INTERFACES IN ASSEMBLIES

The defining characteristic of a structural assembly is its joints. These are interfaces or mating surfaces which are maintained in contact. Thus, in analyzing the dynamic properties of a structural assembly, it is important to consider not only the component materials but also the dynamic response and the energy dissipation caused by interface effects. Three types of interfaces are important in the dynamic analysis in structural mechanics: dry interface surfaces, lubricated interface surfaces, and adhesive-bonded interfaces.

Unjoined interfaces are generally subjected to a large variety of loads and displacements in service. For the case of dry interfaces (metal-to-metal contact), Coulomb friction provides an important mechanism for dissipating energy under cyclic shear displacement.

Textile assemblies (yarns, woven fabrics, knitted fabrics, and nonwoven structures) and fiber-reinforced plastics have extremely large interfaces between fibers and matrix. This is because the fiber has a large surface area compared to its volume. The characteristics of the interface have a strong effect on the dynamic mechanical properties of these polymer assemblies. Thus, dynamic analyses of twisted yarn and nonwoven fabrics are good examples for exhibiting the effects of the interfaces on mechanical properties.

(1) Geometric effects of twisted yarn. Studies in which the dynamic modulus is measured by sonic-velocity techniques have dealt almost exclusively with fibers, films, or fibrous sheets [3]. A few investigations of yarn assemblies have been reported [4-6]. One study, by Platt, deals with the effect of geometric twist on modulus of elasticity [5]. He was concerned with the observed reduction in the dynamic modulus with increasing yarn twist. He concluded that the decrease in modulus is related to the stress distribution in the yarn generated by the yarn geometry and that the effect can be predicted from the stress analysis of continuous-filament yarns. No attempt was made to establish the mechanism of wave propagation in the structure or to show why the static and dynamic behavior are so related.

Zorowski and Murayama reported the results of theoretical and experimental studies of the effect of twist geometry on the effective dynamic modulus of a continuous-filament yarn in both the linear and the nonlinear stress-strain region [6,7]. This investigation was initiated to establish how small-amplitude strain waves, such as occur in the pulse-propagation measurement of the dynamic modulus, are transmitted longitudinally in a continuous-filament twisted yarn.

A model is proposed in which the viscoelastic behavior of a material is considered negligible due to the rapid rate of strain propagation at sonic velocities. Based on an analysis which assumes plane-wave propagation in a continuous medium, it is demonstrated that the effective dynamic modulus and static-extension modulus of a continuous filament yarn are reduced in the same manner by increasing geometric twist. Detailed mathematical analyses of wave propagation in continuous-filament twisted yarn are given in their articles [6,7]. An effective dynamic modulus E_e for the twisted yarn in the linear stress-strain region is expressed by

$$E_e = \frac{2E}{R^2} \int_0^R f(\gamma) \cos^2 \alpha\gamma \, d\gamma, \tag{4.1}$$

so that the propagation velocity for a plane wave becomes

$$C = \sqrt{\frac{E_e}{\rho}}, \qquad E_e = C^2 \rho, \tag{4.2}$$

where E is the extension modulus, R is the yarn radius, $f(\gamma)$ is the radial strain function, α is the filament helix angle at radius γ, γ is the radius to any point in the yarn cross-section, C is the sonic velocity, and ρ is the mass density of the material.

The following expression was developed to predict the dynamic modulus of twisted yarn in the nonlinear region by combining the wave equation with the twist geometry [7]:

$$\frac{E_e}{E_c} = \frac{1}{u} \left[\frac{\varepsilon_y}{\varepsilon_c}\right]^m \frac{2}{3(3 + m)} \left\{1 - \left(1 - \frac{3}{2} u\right)^{3+m}\right\}, \tag{4.3}$$

where u is tan α_{R0} (α_{R0} being the helix angle of the yarn), E_c is the effective dynamic modulus of the yarn in the low-strain region, ε_c is the yield point, ε_y is a given strain in the nonlinear region, and m is the slope of the modulus-strain curve on a semilog scale.

Application of these effective-modulus relationships requires a knowledge of the yarn geometry ($u = \tan \alpha_{R0}$) and filament-material properties (E_c, m, ε_c). Once the strain ε_y to which the yarn is initially subjected is specified, it is first necessary to determine whether the response will be linear, transitional, or nonlinear. This is done by determining what interval ε_y falls in and making use of the appropriate effective-modulus equation [7]. For example, in the strain interval $0 \leq \varepsilon_y \leq \varepsilon_c$ (linear), equation (4.3) becomes

$$\frac{E_e}{E} = \frac{1}{4} + \frac{9}{4}\,\frac{1}{1 + u} - \frac{3}{2u}\,\ell n(1 + u). \qquad (4.4)$$

This equation is identical to that developed by Hearle [8] and Treloar [9] in their static-extension analysis of continuous-filament twisted yarns. The effect of twist geometry on the dynamic modulus is thus predicted to be the same as its effect on the static modulus in the linear region.

Figure 4.1 shows the effective dynamic modulus of hand-twisted continuous-filament yarns of various materials as a function of twist

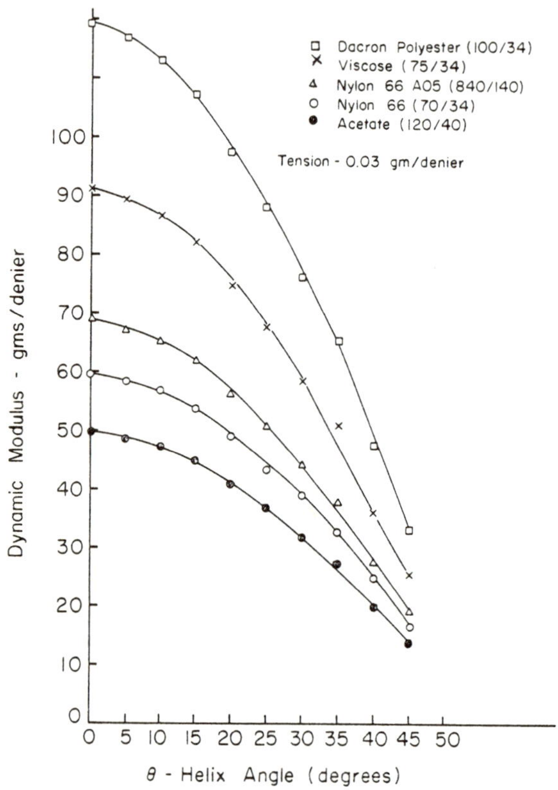

Fig. 4.1. Measured dynamic-modulus variation with yarn twist.

angle, measured by the pulse-propagation technique. These consisted of nylon 66 (70/34) (i.e., a yarn composed of 34 filaments with a total denier of 70), dacron polyester (100/34), acetate (120/40), viscose (75/134), nylon 66 A05 (840/140), high-tenacity viscose (840/560), and polyester (100/34).

All test samples were prepared by hand-twisting the individual yarns on an Alfred Suter twist counter. A small tension (0.03 g/denier) was placed on each sample during twisting, and the yarn was permitted to contract as the helix angle increased. A correction was made for the contraction effect on the helix angle [6].

The effective dynamic modulus was measured on the twisted sample without removing it from the twist counter. The PPM-5 Dynamic Modulus Tester, modified to record flight time over a given specimen length, was used to make the measurements. The PPM-5 transducers were mounted on the twist-counter frame directly below and in contact with the test specimen at separations of 20 and 40 cm. Knowing the time required for the high-frequency strain pulses generated by the instrument to travel a specified distance, the wave propagation velocity could be computed. With the density of the sample known, the effective dynamic modulus was calculated from equation (4.2).

Measurements of dynamic modulus were made on each sample at twist-angle increments of 5°, starting at zero, increasing to 45°, and then decreasing to zero again. The results did not depend significantly on the sample length or on whether at a specific helix angle it was being increased or decreased. The dynamic modulus under low initial yarn tension decreases with increasing helix angle.

Figure 4.2 shows the effect of both twist level and yarn strain on the dynamic modulus of viscose and acetate twisted yarn. To change the yarn strain at a fixed twist level, one end of the yarn was displaced longitudinally in increments of 0.5% strain up to about 8% total strain. Flight-time measurements were made as the strain was applied. The resulting flight times were used to determine the dynamic modulus as indicated previously.

Based on the yarn dynamic-modulus measurements at low initial yarn tension, an effective dynamic-modulus ratio E_e/E, normalized with respect to the zero-twist modulus, was determined [7]. This ratio is in excellent agreement with the variation of modulus ratio with helix angle predicted by equation (4.4). This indicates that the characteristic decrease in the effective dynamic modulus in the linear region is independent of the material or yarn composition (total denier, number of filaments, or denier per filament) and is principally a function of yarn twist geometry.

Using the data from Fig. 4.2 for the variation of effective dynamic modulus with yarn strain at zero twist (i.e., M, ε_c), equations (4.3)

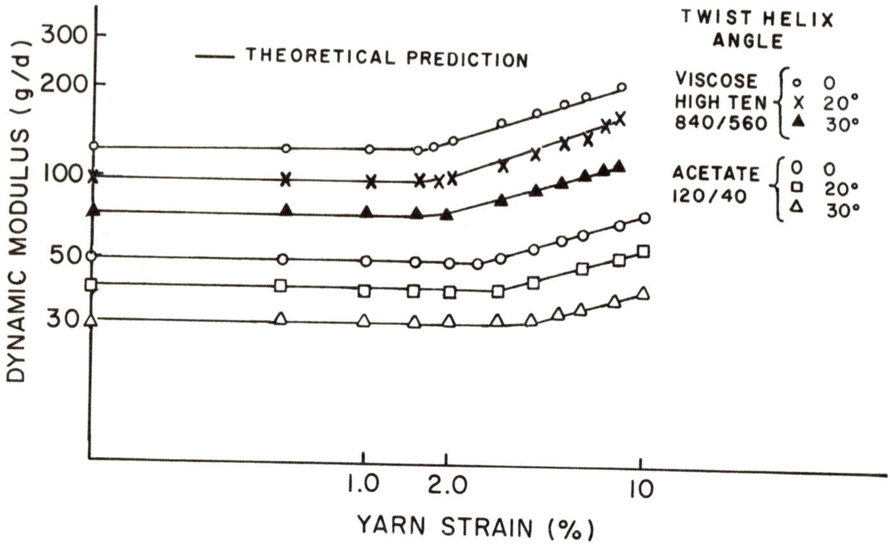

Fig. 4.2. Viscose (high tenacity) and acetate.

and (4.4) can be applied to determine how this variation is affected by geometric twist in the transitional and nonlinear regions. The results are shown in Fig. 4.2 for the acetate and viscose yarn. It is seen that the agreement between experiment and theory is excellent up to yarn strains of 8-10%.

From this agreement between the measurement and theory, the following conclusions are drawn concerning high-frequency periodic wave propagation in twisted yarn composed of continuous viscoelastic filaments.

(a) When subjected to a high-frequency periodic strain variation, the viscoelastic filaments from which the textile yarn assembly is composed behave like a crystalline solid with a nonlinear stress-strain curve.

(b) Under the conditions of loading investigated, the variation of dynamic modulus with twist angle in the linear region of filament behavior is principally a geometric effect and is predictable. In a similar fashion, the effects of both changes in filament mechanical properties and those of twist can be accounted for quantitatively when the initial yarn extension is sufficient to initiate nonlinear filament behavior.

(c) In both the linear and the nonlinear region, the high-frequency strain waves generated by the sonic-velocity pulse-propagation method of making measurements are transmitted along a twisted continuous-filament yarn principally as one-dimensional longitudinal plane waves.

These studies were conducted using a sonic-velocity pulse-propagation measurement technique in which viscous dissipation was small and could be neglected. The pulse-propagation method uses a 60-pulse/sec transmitting transducer having a displacement of 0.025 mm. Each stress pulse is allowed to dampen completely before another is introduced, and the only measurement required is the time taken for a pulse to travel a measured distance in the specimen. Since twist yarn exhibits only a small strain amplitude, the internal fiber-to-fiber friction is negligible [10].

There is no significant difference in the modulus values obtained from the yarn and from its constituent fibers, since each filament provides an unbroken path for the stress pulse from the transmitter to the receiver, and since there is no transfer of the pulse from fiber to fiber through frictional contacts.

The study of strain waves in twisted viscoelastic filament assemblies has been extended into the region of low-frequency vibration where dissipative processes, both viscous and Coulomb, can no longer be considered negligible [11,12]. Extensive mathematical analyses have been made on a model which includes a dissipative mechanism allowing for Coulomb damping. This Coulomb (dry friction) damping results from relative movement between the filaments when the assembly undergoes a dynamic response.

In the analysis that follows the technique of <u>equivalent viscous damping</u> is employed. An <u>equivalent viscosity</u> for the material is calculated which will result in identical energy dissipation under identical sinusoidal strain input. The preceding development can then be readily applied by defining an <u>effective viscosity</u> by adding the <u>equivalent viscosity</u> η_{eq}, to the actual viscosity η^*:

$$\eta_c = \eta^* + \eta_{eq}. \tag{4.5}$$

Similarly,

$$\tan \delta = \tau\omega + \frac{\eta_{eq}\omega}{E_e}, \tag{4.6}$$

where δ is the phase angle between stress and strain, τ is the relaxation or retardation time, ω is the frequency of vibration, and E_e is the effective assembly dynamic modulus.

The magnitude of the complex modulus ($|E|$) and the loss tangent ($\tan\delta$) of a continuous-filament twisted yarn are related to its structural geometry and filament properties by

$$|E| = \frac{EI}{\cos\delta}, \quad \tan\delta = \eta\omega + \eta_{eq}\frac{\omega}{E}, \tag{4.7}$$

where I and η_{eq} are defined as follows. If

$$E_e \frac{\partial^2\mu}{\partial X^2} + \eta^* \frac{\partial^3\mu}{\partial X^2 \partial t} = \rho \frac{\partial^2\mu}{\partial t^2}, \tag{4.8}$$

where $E_e = EI$, $\eta^* = \eta I$, with

$$I = 2\int_o^1 \left(\frac{\gamma}{R}\right) f\left(\frac{\gamma}{R}\right) \cos^2\theta \, \frac{\gamma}{R} \, d\left(\frac{\gamma}{R}\right),$$

$$f\left(\frac{\gamma}{R}\right) = \left.\frac{\partial\phi}{\partial\varepsilon_y}\right|_{\varepsilon_y=\varepsilon_s}.$$

This is the governing equation of motion for the twisted model yarn in vibration. For η_{eq} we have

$$\eta_{eq}\frac{\omega}{E} = \frac{6\mu(2\varepsilon_s/\varepsilon_o + 1)}{k}\phi(\alpha), \tag{4.9}$$

where

$$\phi(\alpha) = \frac{\tan\alpha}{2\,\ell n\,(1 + \tan^2\alpha) - \tan^2\alpha}\left\{\frac{3}{2}\frac{\alpha}{\tan\alpha} - \frac{1}{2} - \frac{1}{1 + \tan^2\alpha}\right\}.$$

Values of the function $\phi(\alpha)$ are given in Table 4.1. In these equations,

TABLE 4.1
Variation of $\phi(\alpha)$ with

α	0°	10°	15°	20°	25°	30°	35°	40°	45
ϕ	0.00	0.0881	0.1373	0.1836	0.2300	0.2858	0.3128	0.3768	0.46

E_e = effective assembly dynamic modulus,
μ = longitudinal displacement of yarn axis,
X = position along yarn axis,
t = time,
η^* = assembly internal viscosity,
ρ = material mass density
γ = radial position in yarn cross-section,
R = yarn radius,
ε_y = yarn strain,

ε_s = initial static yarn strain,
ε_o = magnitude of sinusoidal strain input,
α = yarn helix angle.

The dynamic mechanical properties of continuous-filament twisted yarns have been measured by means of low-frequency vibration at different twist levels, using a Rheovibron. The samples consisted of a nylon-66-A05 tire yarn with 140 filaments and a total denier of 840 (840/140) and two textile yarns: polyester (100/34) and acetate (120/40). In fig. 4.3 are plotted the real part of the dynamic

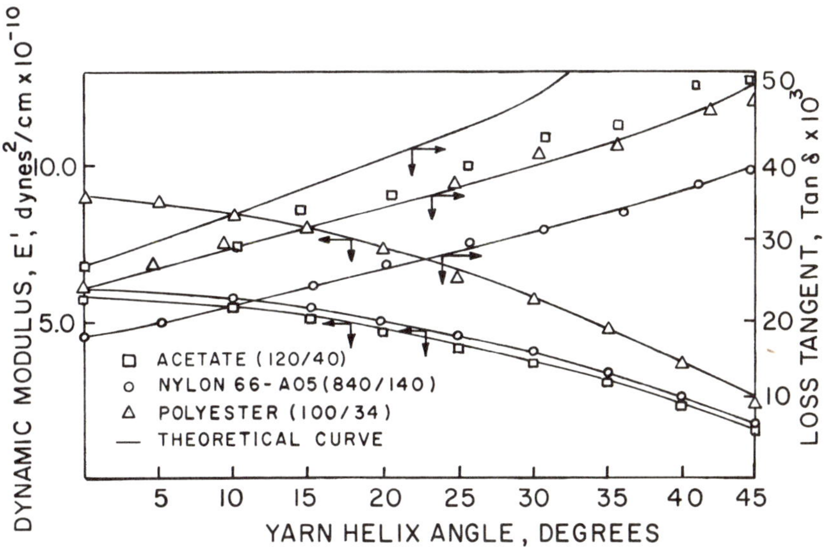

Fig. 4.3. Measured and predicted effect of twist geometry on dynamic modulus and loss tangent for nylon, polyester, and acetate yarns.

modulus (E') and the loss tangent (tan δ) for these materials as determined from Rheovibron data at 110 Hz, as a function of the yarn helix angle α. The dynamic mechanical properties at 110 Hz as a function of the yarn twist level, predicted by using the equations with the experimental values of the integral I and the proportionality constant k [11], are plotted as solid curves in Fig. 4.3. The loss tangent (tan δ) increases with increasing yarn helix angle in twisted yarn. This result indicates that under conditions of low initial static strain, a dissipative mechanism in addition to internal material viscous damping must be introduced to account for the total energy dissipation in the twisted assembly. This appears to be explainable in terms of the energy dissipated by Coulomb friction between filaments which move relative to each other due to variation in the local helix angle as the strain wave propagates along the yarn.

(2) Assembly factors in nonwoven fabrics. The characteristics of interfaces between filaments have a large effect on the dynamic mechanical properties of textile assemblies (yarns, woven and knitted fabrics, and nonwoven structures). Dynamic mechanical analysis permits the quantitative characterization of filament-to-filament friction and bonding strength in nonwoven fabrics using the loss tangent tan δ [13].

This procedure is derived from a phenomenological model of the structure's mechanical behavior in which its energy-dissipation capacity consists of both filament internal viscous friction and filament-to-filament Coulomb friction. By treating the Coulomb dissipation in terms of an additional effective viscous component, the loss tangent of the model representing the structure can be expressed analytically in terms of the internal filament viscous properties, a fiber-to-fiber Coulomb parameter, and an undetermined assembly dispersion factor representing the bond strength. By measuring the effective loss tangent for the structure and for a constituent filament at two separate frequencies, and considering the magnitude of the Coulomb mechanism to be independent of the test frequency in the range employed, both the assembly dispersion factor and the Coulomb-friction dissipation can be calculated.

The dynamic mechanical test procedure consists of applying a sinusoidal displacement to a fabric sample nondestructively and monitoring the output response. The magnitude and phase change of the output signal can be related to certain effective mechanical properties of the sample. To analyze dynamic data, the simplest useful model representing a fabric structure consists of three lumped mechanical elements as depicted in Fig. 4.4 (a), i.e.,

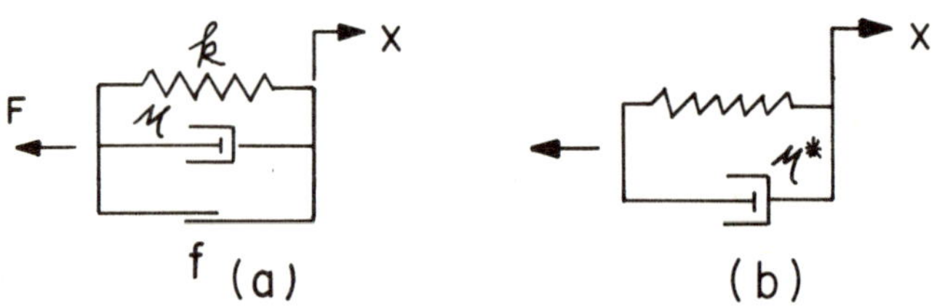

Fig. 4.4. Mechanical model.

k, a linear elastic spring which stores energy,

η, an energy dissipation mechanism associated with the internal viscosity of the filaments,

f, an energy dissipation mechanism of the Coulomb form associated with filament-to-filament surface friction.

Coulomb friction, which is nonlinear, complicates the analysis of the model. For simplicity it is replaced by an equivalent viscosity for which the energy dissipated per cycle is identical to that of the actual Coulomb phenomenon. This additional pseudoviscous effect can be combined directly with the internal filament losses in an effective viscosity η^* as depicted in Fig. 4.4 (b). Mathematically the equivalent Coulomb viscosity can be represented as

$$\eta_{eq} = \frac{4f}{A\omega}. \qquad (4.10)$$

The total effective viscosity then becomes

$$\eta^* = \eta + \frac{4f}{A\omega}. \qquad (4.11)$$

In terms of the model depicted in Fig. 4.4 (b), the force output from a sample subject to a sinusoidal input can be written

$$F = kx + \eta^* \frac{dx}{dt}, \qquad (4.12)$$

where

$$x = A \sin \omega t. \qquad (4.13)$$

Combining (4.12) and (4.13) and rearranging permits the force output to be expressed as

$$F = \frac{kA}{\cos \delta} \sin(\omega t + \delta), \qquad (4.14)$$

where

$$\tan \delta = \frac{\eta^* \omega}{k}. \qquad (4.15)$$

Here k is the elasticity of the fabric model, f is its Coulomb friction coefficient, η is its filament viscosity, η_{eq} is the equivalent viscosity representing Coulomb friction losses, A is the amplitude of the imposed oscillation and ω its frequency, η^* is the total effective viscosity of the fabric model, F is its force output, x is the input oscillation displacement, t is time, and δ is the phase shift between output force and input displacement.

From (4.14) and (4.15) it is seen that the phase shift of the sample response, expressed as tan δ, is a direct measure of the ratio of the energy dissipated to the energy stored. This is a particularly convenient parameter to work with, since it is measured directl by the Rheovibron.

By separating the loss tangent into its two constituent parts (internal filament viscosity and filament-to-filament surface friction), parameters can be defined which can be interpreted as being related to the bond strength. The effective loss tangent for a fabric sample measured at a specific frequency can be written in the form

$$\tan\delta\Bigg|_{\omega_i}^{fab} = C\,\frac{\eta_f\omega_i}{k}\Bigg|_{\omega_i} + \frac{4f}{Ak}. \qquad (4.16)$$

The factor C will be dependent on both fabric geometry and bond strength. For given geometry, C will increase with increasing bond strength, approaching 1 in an ideal, completely bonded structure. This factor can also be interpreted as a measure of assembly dispersion, since it represents the facility with which the structure transmits a mechanical input, and it involves the individual filaments dissipating energy through internal viscosity.

The quantity 4f/Ak is very simply related to the interfilament friction and as such is indirectly representative of the degree of bonding in a given structure. This parameter is expected to increas as C decreases. It is not anticipated, however, that these two parameters are related by any simple functional form.

If C and f are assumed to be independent of frequency (and arguments can be made to justify this assumption), then measurements at a second frequency can be combined with the formulation in (4.16), recognizing that

$$\tan\delta\Bigg|_{\omega_i}^{fil} = \frac{\eta_f\omega_i}{k}\Bigg|_{\omega_i}, \quad i = 1 \text{ or } 2, \qquad (4.17)$$

to give for C and 4f/Ak the following expressions:

$$C = \frac{\tan\delta\Big|_{\omega_1}^{fab} - \tan\delta\Big|_{\omega_2}^{fab}}{\tan\delta\Big|_{\omega_1}^{fil} - \tan\delta\Big|_{\omega_2}^{fil}} \qquad (4.18)$$

$$\frac{4f}{Ak} = \tan\delta \Bigg|_{\omega_1}^{fab} - C \tan\delta \Bigg|_{\omega_1}^{fil} , \qquad (4.19)$$

where ω_1, ω_2 are the chosen test frequencies, M_f is the filament viscosity, and the superscripts are fab for fabric, fil for filament.

It is observed that all terms on the right side of (4.18) and (4.19), with the exception of C, can be measured directly on the Rheovibron viscoelastometer using actual fabric and constituent fiber samples.

A series of loss-tangent measurements were made on three samples of spun bonded nonwoven fabric in both the machine and transverse directions using the Rheovibron. The samples chosen were representative of fabrics with good, intermediate, and poor bond strength. These tests were conducted at four frequencies as reported in Table 4.2. Equations (4.18) and (4.19) were used to compute the bonding

TABLE 4.2

Loss Tangent ($\tan\delta$) of Spun Bonded Nonwoven Fabrics

		$10^{-3}\tan\delta$			
Sample	Meas.[a]	3.5 Hz	11 Hz	35 Hz	110 Hz
A	MD	41.2	40.5	36.2	35.2
	TD	44.2	43.2	40.1	39.3
	F	32.5	30.2	26.5	24.1
B	MD	58.2	56.3	52.2	50.0
	TD	64.3	62.2	59.1	57.3
	F	38.3	34.2	26.3	22.4
C	MD	77.3	74.1	71.0	68.2
	TD	82.2	80.4	77.2	75.1
	F	58.3	44.1	30.4	15.4

[a]MD, machine direction; TD, transverse direction; F, component fiber.

and friction-factor parameters quantitatively for all combinations of frequency with a ratio of 10 or greater. These data are presented in Table 4.3 along with tenacity measurements made statically on the same fabric with an Instron. These tenacity tests were performed on samples with a gauge length five times the width. The correlation between the bonding factor C (measured nondestructively) and the tenacity is striking. The smallness of the variation of both C and 4f/Ak with test frequency represents further evidence that they represent geometric characteristics of the fabric structure related to bond strength and filament-surface interaction.

TABLE 4.3

Bond strength parameter (C) and interfilament friction factor (4F/Ak) of spun-bonded nonwoven fabrics

Sample	Meas.[a]	C 11-110	C 3.5-35	C 3.5-110 Hz	4F/Ak 11-110	4F/Ak 3.5-35	4F/Ak 3.5-110 Hz	Instron tenacity lb/in. oz/yd²
A	MD	.832	.826	.752	.020	.019	.018	19.2
	TD	.666	.660	.625	.027	.026	.029	13.5
B	MD	.512	.505	.520	.045	.044	.047	12.6
	TD	.415	.420	.432	.053	.053	.054	7.2
C	MD	.208	.213	.209	.074	.071	.071	5.8
	TD	.173	.178	.164	.078	.077	.079	1.8

[a]MD, machine direction; TD, transverse direction.

3. ADHESION

In a study [14] of the bonding in reinforced composites it was found that energy is dissipated at the filament-matrix interface when the composite is subjected to continuous cyclic loading if adhesion is poor. Also, a theory has been developed which permits the separation of the total energy dissipation in a composite system under cyclic loading into a portion associated with the viscous properties of the constituent materials themselves and one resulting from the lack of perfect adhesion at the filament-matrix interface. This provides a mechanism for quantitatively characterizing the degree of adhesion by means of a dynamic test using instruments like the Rheovibron on composite systems of a variety of materials.

This dynamic energy-loss measurement for characterizing adhesion should apply to other composite systems. Thus the tire cord-rubber system has also been studied by this procedure, and the relationships between these dynamic measurements and static adhesion test (H-block pull test [15] and strip-peeling test [16]) have been reported [17].

(1) Dynamic adhesion test by energy-loss measurements. The Rheovibron dynamic viscoelastometer was used for this investigation. The principle of this direct-reading method [18] using the Rheovibron instrument was extended to the dynamic mechanical analysis of the two-component system for determining the energy dissipation (tan δ) due to poor adhesion. The details of these analyses are given by Zorowski and Murayama [14], who obtain the following relations:

$$\tan \delta_{adh} = \tan \delta_{exp} - \tan \delta_s, \qquad (4.20)$$

$$\tan \delta_s = \frac{\tan \delta_f \, E_f \, V_f + \tan \delta_m \, E_m \, V_m}{E_m \, V_m + E_f \, V_f}, \qquad (4.21)$$

where $\tan \delta_{adh}$ is the internal energy dissipation due to poor adhesion, $\tan \delta_{exp}$ is the measured effective loss tangent for the composite (cord-rubber), $\tan \delta_s$ is the effective loss tangent for a composite with perfect adhesion, $\tan \delta_f$ is the filament (cord) loss tangent, E_f is the filament modulus, V_f is the filament volume fraction, $\tan \delta_m$ is the matrix (rubber) loss tangent, E_m is the matrix modulus, and V_m is the matrix volume fraction.

By measuring the total system energy dissipation in terms of tan δ and knowing tan δ and dynamic moduli of the components as well as the volume-fraction composition, the dissipation due to the poor interface adhesion can be determined. Dynamic measurements of the internal energy dissipation of adhesive-dipped tire cords and of tire-cord-rubber blocks, 0.5 X 5 cm and about 0.1 cm thick were made at a frequency of 11 Hz with a strain amplitude of 0.5%. Samples were heated at 1°C/min in a nitrogen atmosphere in a relaxed condition. Measurements of the tensile modulus E, and the loss tangent tan δ were made at 10 or 20°C increments. Samples were allowed to equilibrate at temperature for 15 minutes before measurements were made.

In order to examine the relation between dynamic energy-loss measurements and the H-block pull test or the strip-peeling test, portions of H-blocks and strips were utilized in the dynamic test. The commercial tire cords were nylon 66 twisted into a 2-ply, 840-denier/ply cord with 12 turns per inch Z-twist in singles and 12 turns per inch S-twist in the ply (6 denier per filament), and polyethylene terephthalate (PET) twisted into a 2-ply, 1000-denier/ply cord with 12 turns per inch Z-twist in singles and 12 turns per inch S-twist in the ply (5 denier per filament). The cord was dipped through a bath of adhesive and passed through a drying oven at 165°C and then into an electrically heated circulating hot-air oven for 1- to 3-minute cures at constant length. The dry add-on of adhesive was approximately 5%.

The cured cord was tested for static adhesion at 120°C in the H-pull test as described in ASTM D-2138-67 with the exception that 3/16 in. rather than 1/4 in. of cord was pulled from rubber [15]. The peeling adhesion of 1 X 6-in. strips of a cord-rubber composite were tested at 120°C. The construction of these strip adhesion samples was described by Brownlee [16].

The two plies of cord are parallel. The strips were pulled at 12 in./min at 180° angle, and the peel force was averaged over a 2-in. length of peel. The plies were built with the closest possible spacing of cords. The thickness of the skim stock between plies was 15 to 20 mils.

The rubber used in H-blocks and strips was a commercial skim stock. The Mooney scorch time (5-point rise at 135°C) was 13.5 min; a Monsanto rheograph indicated a cure time of 20 min at 153°C. The H-block and strip-pad mold surfaces were subjected to 500 lb/in.^{2} in a Sadena hydraulic press, Model Q-230, at 300°F for 25 min during vulcanization.

Pull loads listed in the tables are average values from eight H-blocks. Results are stated as 95% confidence intervals. Peeling loads are average values from five strips from a single mold. The strip-peeling loads exhibited a maximum average deviation from the mean of $\pm$ 2 lb.

The H-7 dip consisted of 82 ml of H-7 adhesive (Imperial Chemical Industries Ltd.), 50 ml of Gen-Tac Latex (General Tire & Rubber Company, polyvinylpyridine-styrene-butadiene latex, 41 wt % solids), and 73 ml of deionized water.

A standard RFL dip was prepared by dissolving 5.5 g of resorcinol in 116 ml of water containing 0.15 g of sodium hydroxide. To this solution, 8.1 g of 37 wt % aqueous formaldehyde was added, and the solution was stirred for 5 min. This resin master solution was allowed to age for 6 hr at 75-78°F. The 133 g of resin master was added to a dispersion of 122.0 g of Gen-Tac and 30.0 g of deionized water and the resulting dispersion was mixed for 15 min. The RFL dip was used within 72 hr.

Cords used in H-blocks dipped in the H-7 adhesive were cured at 230°C for 3 min, and those in the RFL adhesive at 220°C for 1 min. The PET cords utilized in strip tests were dipped in RFL and cured at 230°C for 3 min; the nylon 66 cord dipped in RFL was cured at 210°C for 3 min.

Two of the PET cords used in strip testing had received surface treatments in electrical discharges to increase their adhesion to RFL. These surface treatments did not alter the bulk properties of these two cords.

The composite blocks for dynamic testing (0.5 X 5 cm and 0.1 cm thick) were cut from static test samples. The composite from the H-block sample contained a single centered cord composing 4 to 6% volume fraction of the sample. The composite from the strip sample contained seven ends of parallel cords composing 28 to 32% volume fraction of the sample.

(2) Results of adhesion tests. The dynamic modulus (E_m) and the internal friction ($\tan \delta_m$) of the vulcanized skim-stock rubber used for test samples are shown in Fig. 4.5. The dynamic moduli (E_f) and loss tangent ($\tan \delta_f$) of the dipped cords (nylon 66 RFL 210°C and PET H-7 230°C) as a function of temperature at 0% RH are shown in Fig. 4.6.

The internal friction of both dipped textile cords show the α-transition peak, which is connected with the glass transition, in which polymer chain segments acquire considerable mobility. The values of tan δ were higher than those in yarns in the temperature range from 25°C to 100°C [19]. This difference indicates that the twist geometry and the surface coating of adhesive on the cord influenced the energy dissipation in the system. Figure 4.7 shows the loss tangent ($\tan \delta_{exp}$) of dipped cord-rubber samples from H-blocks

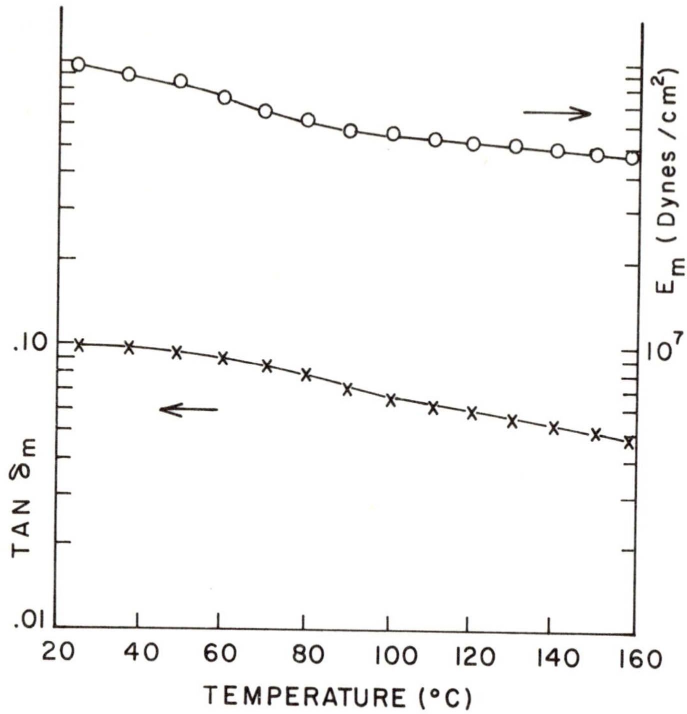

Fig. 4.5. Dynamic modulus (E_m) and internal friction ($\tan \delta_m$) of vulcanized skim stock rubber.

as a function of temperature at 0% RH. The internal friction of dipped nylon 66 (RFL 220°C) or PET (RFL 230°C) cord-rubber composites exhibit the α-transition peak. The tan δ values of these samples were higher than those of the dipped cords. These differences reflect the energy dissipation of rubber and the degree of adhesion. The measured values of the system damping, $\tan \delta_{exp}$, for these dipped cord-rubber samples at room temperature are listed in Table 4.4, along with the system dynamic modulus E_c; the modulus of the dipped cord, E_f; the loss tangent of the dipped cord, $\tan \delta_f$; the energy dissipation with perfect adhesion, $\tan \delta_s$; and the interface dissipation due to lack of perfect adhesion, $\tan \delta_{adh}$.

The results of the H-block pull tests are also listed in Table 4.4. The adhesion factors ($\tan \delta_{adh}$) of nylon 66 and PET tire cords without adhesives were very high ($50.2\text{-}57.0 \times 10^{-3}$), and the pull loads of these samples were low (7.8-9.5 lb). These higher values

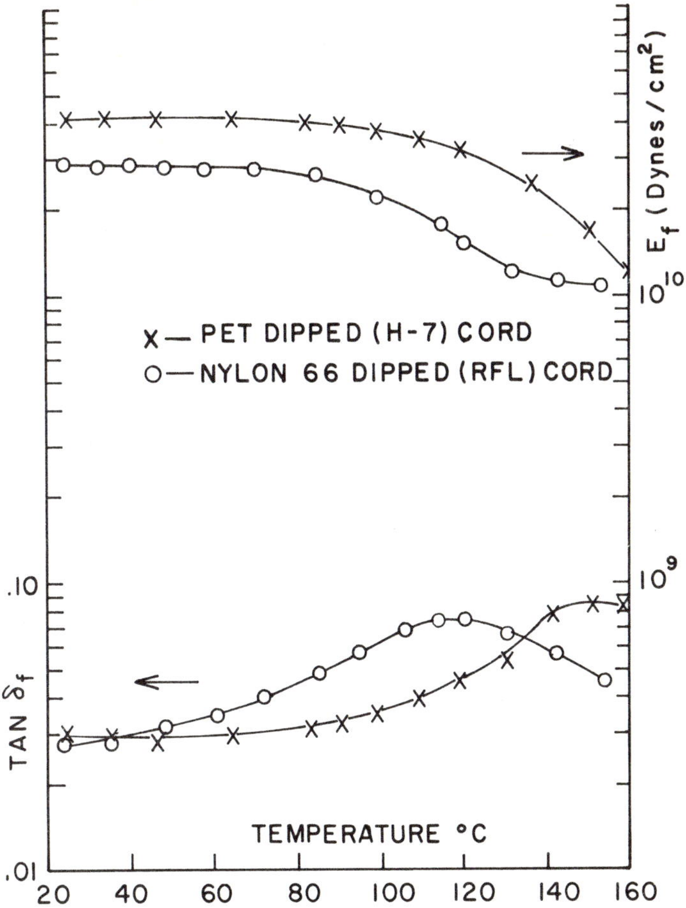

Fig. 4.6. Dynamic moduli (E_f) and loss tangent (tan δ_f) of nylon 66 and PET dipped cords.

of tan δ_{adh} are associated with the bonding of cord to rubber without an adhesive.

It should be noted that for the nylon 66 and PET cords bound to rubber by RFL, the H-block pull loads were very similar: 17.3 and 16.9 lb, respectively. However, the tan δ_{adh} value for the nylon sample was much smaller than for the PET sample. This difference may be related to the unsatisfactory performance of RFL adhesive in PET-reinforced tire building [20]. The static H-block and strip test may not be particularly sensitive to the rheological properties of the bonded interface, while the dynamic energy-loss measurement may potentially yield information on the rheological characteristics

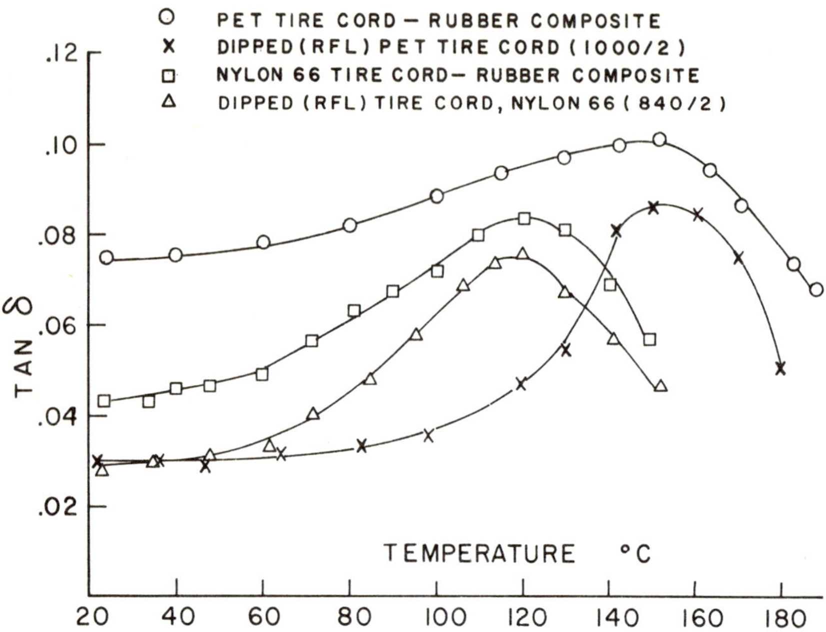

Fig. 4.7. Loss tangent (tan δ_{exp}) of dipped cord-rubber composite (nylon 66, PET).

of the interfacial region of bonding. Kaelble has proposed that the bonding state of an adhesive can be rheologically distinguished from the holding state [21].

In Fig. 4.8, the similar response of the adhesion factor (tan δ_{adh}) and H-block pull load of nylon 66 and PET cord-rubber composites to temperature is shown. The degree of adhesion decreased with increasing temperature to similar extents in the two measurements. There appears to be a reasonable correlation between the two tests.

The peeling loads in the strip adhesion test and the dynamic mechanical properties of four different tire cords and cord-rubber composites are summarized in Table 4.5. These properties were measured at 120°C.

From these results the energy dissipation at the tire cord-to-rubber interface (adhesion factor) was obtained and is also listed in Table 4.5. Figure 4.9 illustrates the relationship of the strip-peeling load to the adhesion factor from Vibron tests. As expected, samples with higher peeling loads show lower values of energy dissipation at the interface (tan δ_{adh}). These results indicate an approximately linear correlation between the strip-peeling measurement and

TABLE 4.4

Adhesion of tire cord to rubber[a] at room temperature

Tire Cord	Dynamic moduli		Internal friction			Adhesion factor	H-block pull load
	E_f 10^{10} dyn/cm²	E_c 10^9 dyn/cm²	10^{-3} tan δ_f	10^{-3} tan δ_s	10^{-3} tan δ_{exp}	10^{-3} tan δ_{adh}	(lb)
Nylon 66 (840/2)							
No adhesive	2.15	0.99	55.0	58.5	108.7	50.2	7.8 ± 0.3
H-7 adhesive	2.26	1.23	65.0	68.1	86.3	18.2	23.6 ± 1.3
RFL adhesive	2.13	1.24	65.0	69.2	81.7	12.5	17.3 ± 1.0
PET (1000/2)							
No adhesive	4.25	1.82	25.0	27.2	84.2	57.0	9.5 ± 0.1
H-7 adhesive	4.50	2.15	33.0	36.2	74.4	38.2	21.0 ± 1.8
RFL adhesive	4.42	2.16	32.1	34.7	80.9	46.2	16.9 ± 1.0

[a] $E_m = 9.3 \times 10^7$ dyn/cm², tan $\delta_m = 0.095$, $V_f = 4.6\%$.

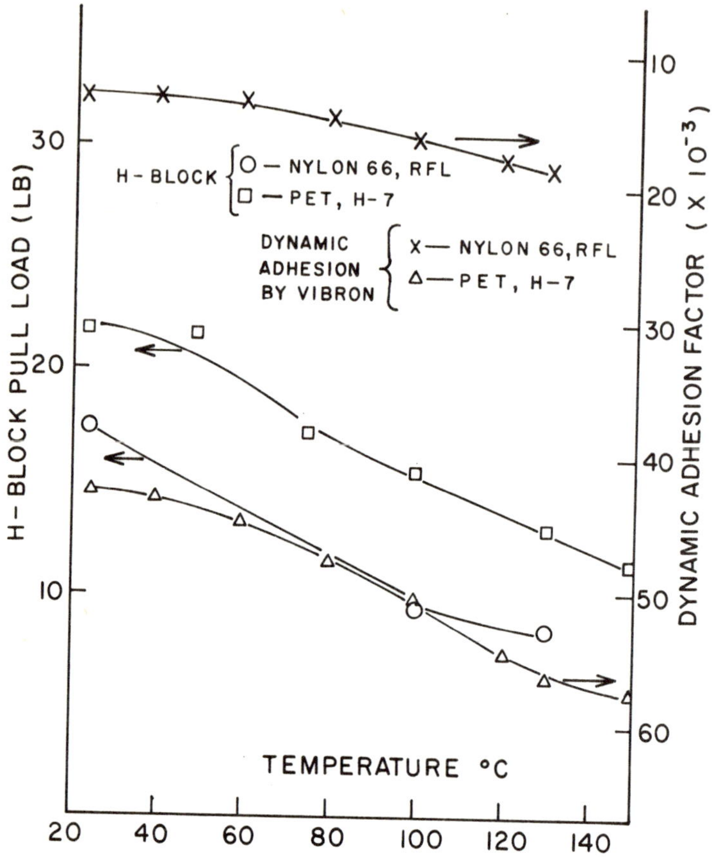

Fig. 4.8. Adhesion factor (tan δ_{adh}) and H-block pull load of nylon 66 and PET cord-rubber composite as function of temperature.

dynamic energy-loss measurement of adhesion, and illustrate the improvement of adhesion of PET to rubber with RFL obtainable by surface modification of the PET.

It has been recognized that static measurements of adhesion between dipped tire cord and rubber may not reflect the performance of the adhesive system under dynamic testing conditions. An adhesive may yield a high initial strip-peeling load, but in dynamic testing under conditions simulating tire running, the bonding of this adhesive may deteriorate more rapidly than that of an adhesive with a lower initial strip-peeling load. The separation of the total energy dissipation in a composite under cyclic loading into a portion

The theory and experimental procedure [14] developed are based on the premise that energy will be dissipated at the matrix-filament interface when the composite is subjected to continuous cyclic loading if bonding is poor. It was postulated that if this interface energy dissipation could be separated theoretically from the other dissipative mechanisms within the composite structure, an instrument for measuring the loss tangent could be used for nondestructive measurement of the level of bonding. The theoretical model which isolates and quantifies these interface losses is based on the work of Feodosev [22], Reshetov [23], and Panovko et al. [24] as reported by Goodman [25]. These studies are theoretical investigations of interface energy dissipation due to slippage in mechanical joints where surfaces are in contact under pressure and the entire system is subjected to cyclic loading.

In the model used to simulate the filament-matrix interface it is assumed that filament slippage may occur due to load and strain differentials developed in the constituent components. Employing a Coulomb friction mechanism to transfer load from the matrix to the filament, the structure is mathematically taken through a complete tension-compression cyclic loading. This results in hysteresis-type force-elongation behavior. The area within the closed hysteresis loop can be interpreted as the energy dissipated at the filament-matrix interface in the model for a complete cycle of loading, as seen in Fig. 4.12 below.

Assuming that these losses will occur in a similar fashion under true dynamic loading, an effective viscosity is defined to simulate the interface dissipation mechanism. This permits the method of analysis developed by Alptekin [26] to be applied to predict the effective loss tangent of the total composite structure. With this formulation, it is possible to separate out the loss tangent due to filament-matrix interface dissipation from experimental data obtained with dynamic instruments.

Figure 4.10 shows the equivalent model used to determine theoretically the energy dissipated under cyclic dynamic loading due to relative movement between a reinforcing filament and its supporting matrix as a consequence of poor adhesion. It consists of a single filament of constant cross-section and properties along its length imbedded in a matrix of lower-modulus material whose cross-section is also constant and whose properties and length are the same as those of the reinforcing filament. Only loadings applied parallel to the filament axis will be considered. For the purpose of the following

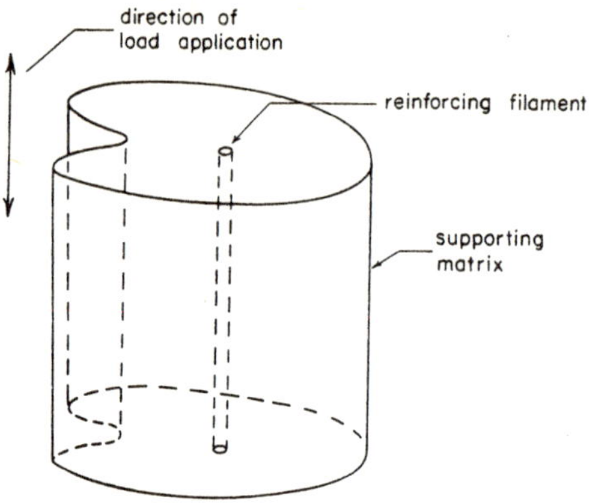

Fig. 4.10. Simulated filament reinforced composite model.

analysis it will be assumed that the cross-sectional area of the matrix (perpendicular to the load direction) is large compared to that of the filament, so that any applied load can be considered to be carried by the matrix surface only. It will also be assumed that the adhesion mechanism between filament and matrix can be simulated by a Coulomb-friction model in which the coefficient of friction along the length of the filament is constant and the normal pressure exerted by the matrix on the filament surface is likewise constant both around its circumference and along its length.

To determine the energy dissipated due to relative movement of the filament with respect to the matrix when the system is subject to cyclic loading, it is first necessary to determine the force-deformation behavior for the composite structure. This will be accomplished by applying a cyclic loading in a pseudostatic fashion and assuming that each component of the system behaves like a linear elastic material. The result of this modeling and analysis for a full cycle of loading will yield hysteresis-type behavior which can be used to determine the energy dissipated due to the simulated loss of adhesion.

The analysis is begun by considering the system, which is initiall unstressed, subjected to an applied tensile loading. Figure 4.11 depicts the internal and external forces acting on the components of the system taken separately as well as the load distribution along their lengths. The load distribution as presented is arrived at in the following manner. Since the load is applied, as previously assumed, only to the matrix at either end of the system, the matrix

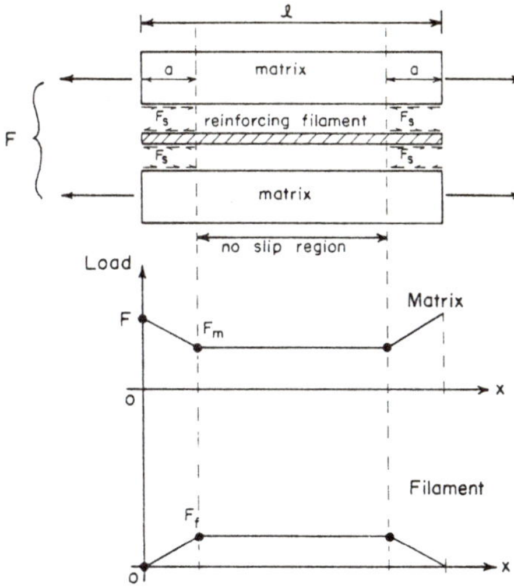

Fig. 4.11. Load distribution and interface slip behavior in composite model under tension.

load is F and the filament load is zero at both x = 0 and ℓ. The matrix load will generate an extensional strain in the matrix. This in turn will give rise to relative movement between the filament and matrix, since there is no strain in the filament. The consequence of this movement will be the generation of a frictional force (due to the Coulomb friction) which will begin producing tension in the filament as indicated. Along with this, there will be a reduction in the matrix tension. This phenomenon will move in from either end until a point is reached where the tensile strains in the matrix and filament are equal and the two extend together without any additional relative movement. This point is defined by the indicated region at each end of the system. Over the central part of the composite the component loads will remain constant. The load will vary linearly over the length as a consequence of the assumption that the pressure between the filament and matrix, as well as the coefficient of friction, is constant. From this load distribution, the force-deformation behavior can be determined by mathematical analysis [14].

Figure 4.12 shows the force-deformation behavior graphically in a plot of the total load F applied to the composite against the parameter $\lambda k_m \delta_s$, where λ is the friction parameter ($\lambda = 2\pi RuP$), K_m is the matrix spring constant, δ_s is the total sample elongation, R

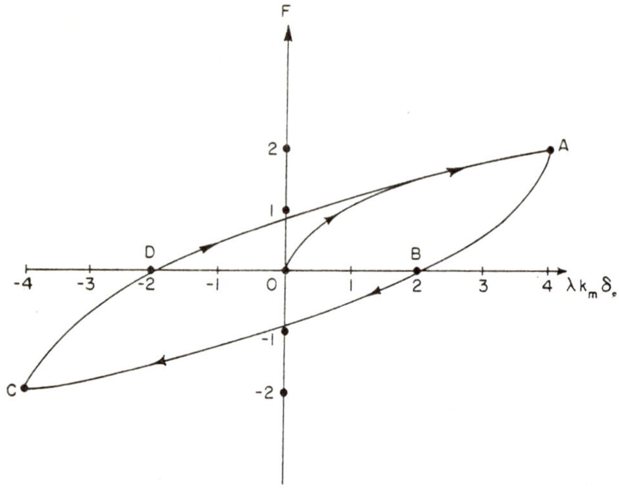

Fig. 4.12. Force-elongation behavior of composite model under cyclic tensile compressing loading.

is the filament radius, u is the interface friction coefficient, and P is the interface pressure.

The effective real and imaginary components of the complex modulus of the system are given by

$$E' = E_e = E_m V_m + E_f V_f \tag{4.22}$$

and

$$E'' = \eta_e \omega = \omega \left[\eta_m V_m + \eta_f V_f + \frac{\eta' n}{A} \right], \tag{4.23}$$

so that

$$\tan \delta = \frac{E''}{E'} = \frac{\eta_e \omega}{E'}. \tag{4.24}$$

Equation (4.22) indicates that the effective dynamic modulus of the system is given by the volume-fraction rule. This is also true for the contributions of the matrix and filament internal viscosities to the imaginary portion of the complex modulus, i.e., if the bonding were perfect, so that η' were zero, then

$$E''_s = \omega(\eta_m V_m + \eta_f V_f). \tag{4.25}$$

Now let $\tan \delta_s$ be defined by

$$\tan \delta_s = \frac{(\eta_m V_m + \eta_f V_f)\omega}{E'} \tag{4.26}$$

to again represent perfect bonding. Then equation (4.24) can be written in the form

$$\tan \delta = \tan \delta_s + \tan \delta_{adh}, \tag{4.27}$$

where

$$\tan \delta_{adh} = \frac{\eta' n\omega}{AE'} \tag{4.28}$$

represents the contribution or influence of the adhesion characteristics on the dissipative properties of the system. A better insight into the role of individual system parameters on this influence is obtained by combining the equations:

$$\tan \delta_{adh} = \frac{4n}{3\pi AE'\beta} (\delta_s^m \lambda k_m)^{1/2} \tag{4.29}$$

This shows that the interface energy dissipation due to lack of perfect adhesion increases in the same fashion with maximum cyclic elongation δ_s^m, matrix spring constant k_m, and interface load-transfer parameter λ, as modeled by the Coulomb friction mechanism. On the other hand, the interface energy dissipation decreases as the effective system dynamic modulus E' increases and as the fiber modulus increases in comparison with the matrix modulus (i.e., $\beta \to 1$). As would be expected, an increase in the number of filaments per unit system cross-sectional area, n/A, also increases the level of interface energy dissipation.

A series of experiments employing the Vibron DDV-II viscoelastometer were designed to verify the proposed theory, based on the following considerations. Consider a composite sample in which the volume fraction of the constituent components can be varied without changing the number of filaments involved or the nature of the bonding between the filaments and surrounding matrix. If such samples are subjected to a constant maximum cyclic deformation, then $\tan \delta_{adh}$ as presented in equation (4.29) can be interpreted and rewritten in the following fashion.

$$\tan \delta_{adh} = \frac{4n(\delta_s^m \lambda)^{1/2}}{3\pi} \; \frac{k_m^{1/2}}{\beta AE'} \; . \tag{4.30}$$

The first factor will not vary as the volume fraction is changed for a given sample and can be represented by a constant characteristic of the level and quality of adhesion in that particular sample, i.e.,

$$\frac{4n(\delta_s^m \lambda)^{1/2}}{3\pi} = C'. \tag{4.31}$$

In terms of volume fractions and constituent-component properties, the remaining terms on the right side of equation (4.30) can be written as

$$\frac{k_m^{1/2}}{A} = \frac{(E_m A_m)^{1/2}}{A} = \frac{E_m^{1/2}}{A_f^{1/2}} \left[\frac{A_m}{A}\right]^{1/2} \left[\frac{A_f}{A}\right]^{1/2} = C'' V_m^{1/2} V_f^{1/2}, \tag{4.32}$$

since the matrix modulus and filament cross-sectional area would remain constant for the samples as defined. Also,

$$E' = E_m V_m + E_f V_f \tag{4.33}$$

and

$$\beta = \frac{k_f}{k_f + k_m} = \frac{A_f E_f}{A_f E_f + A_m E_m} = \frac{1}{1 + A_m E_m / A_f E_f} = \frac{1}{1 + \frac{V_m}{V_f}\frac{E_m}{E_f}}. \tag{4.34}$$

Combining equations (4.29)-(4.33) then results in

$$\tan \delta_{adh} = C \frac{V_m^{1/2} V_f^{1/2}}{E_m V_m + E_f V_f} \left[1 + \frac{V_m E_m}{V_f E_f}\right]. \tag{4.35}$$

This equation describes how the interface energy dissipation varies with the type of experimental sample, in terms of the change in volume fraction of the constituent components, their dynamic moduli, and a constant which characterizes the bonding adhesion between the filaments and the matrix.

By measuring $\tan \delta_{adh}$ experimentally for a series of samples of the type proposed in which the volume fraction is changed without disturbing the interface bonding, the constant C can be evaluated from equation (4.35). If the value of C does in fact remain constant as the volume fraction of the constituent components is varied, the proposed theory is verified.

To determine $\tan \delta_{adh}$ the total effective internal energy dissipation of the composite system is measured as $\tan \delta_{exp}$ using the Rheovibron. Then from equation (4.27)

$$\tan \delta_{adh} = \tan \delta_{exp} - \tan \delta_s, \tag{4.36}$$

where $\tan \delta_s$ from equation (4.26) is given by

$$\tan \delta_s = \left[\frac{\eta_m V_m + \eta_f V_f}{E'}\right] \omega. \tag{4.37}$$

However, recognizing that

$$\tan \delta_f = \frac{\eta_f \omega}{E_f}, \quad \tan \delta_m = \frac{\eta_m \omega}{E_m},$$

and

$$E' = E_m V_m + E_f V_f,$$

equation (4.37) can be rewritten and combined with equation (4.36) to give

$$\tan \delta_{adh} = \tan \delta_{exp} - \frac{(\tan \delta_f)\, E_f V_f + (\tan \delta_m)\, E_m V_m}{E_m V_m + E_f V_f}. \quad (4.38)$$

Thus, by measuring the total effective system energy dissipation in terms of tan δ and knowing tan δ and the dynamic moduli of the constituent components as well as the volume-fraction composition of the sample, the energy dissipation due to the lack of perfect interface adhesion can be determined.

To carry out the experiment as outlined, composite samples were prepared by imbedding single copper wires in an epoxy-828 resin matrix. Two sizes of copper wire were used: 0.25-mm and 0.63-mm diameter. The wire was placed in a rectangular box into which the epoxy resin was poured; MDA was used as a curing agent. The wire was oriented so as to pass longitudinally through the matrix as close to the center of its cross-section as possible. The samples were cured at 150°C for 25 hours. The final dimensions of the sample were fixed by machining the epoxy matrix surface with a precision grinder, taking care to avoid a significant temperature rise in the sample. The original specimens were prepared in this fashion to be 5 cm in length with a rectangular cross-section of approximately 5 X 1 mm. Following measurement of $\tan \delta_{exp}$ for the samples as prepared, the volume fraction was varied by grinding the matrix surface and reducing the cross-sectional dimensions of the samples.

Four different samples of the type described were prepared. These consisted of the two wires of different diameter, each imbedded in two epoxy resins of differing properties. In Table 4.6 are summarized the room-temperature mechanical properties of the components that went into these four samples.

Dynamic measurements of the total effective internal energy dissipation of these composite samples were made on the Rheovibron at a frequency of 11 Hz with an applied average strain amplitude of 0.5%. All tests were conducted at room temperature. The measured values of the system damping, $\tan \delta_{exp}$, for all the model samples are listed

TABLE 4.6

Sample	Diameter (mm)	Copper Wire E_f (10^{10} dyn/cm^2)	tan δ_f	Epoxy E_m (10^{10} dyn/cm^2)	tan δ_f
I	0.25	83	.003	0.5	.042
II	0.63	83	.003	0.5	.042
III	0.25	83	.003	1.0	.030
IV	0.63	83	.003	1.0	.030

in Table 4.7. Also summarized in Table 4.7 are the calculated values of the effective system dynamic modulus E'; the effective internal energy dissipation with perfect adhesion, tan δ_s; the interface intern energy dissipation due to lack of perfect adhesion, tan δ_{adh}; and the adhesion characterization constant C. In Table 4.8 are listed the average values of the constant C for the four samples investigated and the percentage variation in this number for each sample. To provide a better appreciation of the order of magnitude of the effect of the interface energy dissipation due to lack of perfect bonding, the values of tan δ_{adh} and tan δ_s for the four samples are plotted as a function of matrix volume fraction in Fig. 4.13.

An examination of the percentage variation in the mean values of the experimentally determined adhesion constant C as presented in Table 4. would appear to indicate, on the basis of the limited tests conducted, that the theory proposed for filament-matrix interface energy dissipation possesses a reasonable degree of credibility. The percentage variations are all within the range of experimental error for the type of measurement being made, and the similarity in magnitude of the mean values of C is not unreasonable in light of the materials and construction of the experimental samples tested. It thus appears that a theory has been developed which permits the separation of the total energy dissipation in a composite system under cyclic loading into that portion associated with the viscous properties of the constituent materials themselves and that resulting from the lack of perfect adhesion at the filament-matrix interface. This in turn provides a mechanism by which it should be possible to characterize the degree of adhesion quantitatively by means of a relatively simple dynamic test using the Rheovibron or other instruments with similar capabilities on composite systems of a variety of materials.

The results presented in Table 4.7 and Fig. 4.13 indicate that the internal energy dissipation at the filament-matrix interface is from

TABLE 4.7

Experimental and calculated adhesion characterization results

Model	Wire diameter (mm)	V_f	V_m	E' (10^{10} dyn/cm^2)	tan δ_s	tan δ_{exp}	tan δ_{adh}	C (10^{12} dyn/cm^2)
I	.25	.05	.95	4.6	.0070	.0240	.0170	321
		.11	.89	9.05	.0050	.0170	.0120	332
		.18	.82	15.5	.0038	.0140	.0102	397
		.30	.70	24.5	.0035	.0115	.0080	387
II	.63	.05	.95	4.5	.0071	.024	.0169	312
		.14	.86	11.4	.0043	.0155	.0112	355
		.42	.58	34.2	.0033	.0092	.0059	360
III	.25	.05	.95	5.7	.0080	.0210	.0130	278
		.09	.91	9.0	.0059	.0165	.0106	297
		.22	.78	19.5	.0041	.0110	.0068	308
		.36	.64	31.2	.0035	.0085	.0049	310
IV	.63	.107	.893	10.5	.0055	.016	.0105	323
		.18	.82	16.0	.0044	.013	.0086	339
		.30	.70	26.0	.0037	.010	.0063	348
		.45	.55	38.2	.0034	.008	.0046	349

TABLE 4.8
Variation of adhesion constant C

Sample	C[a]	Max. Variation (%)
I	359	10.6
II	343	9.1
III	298	6.7
IV	339	4.7

[a]Mean value.

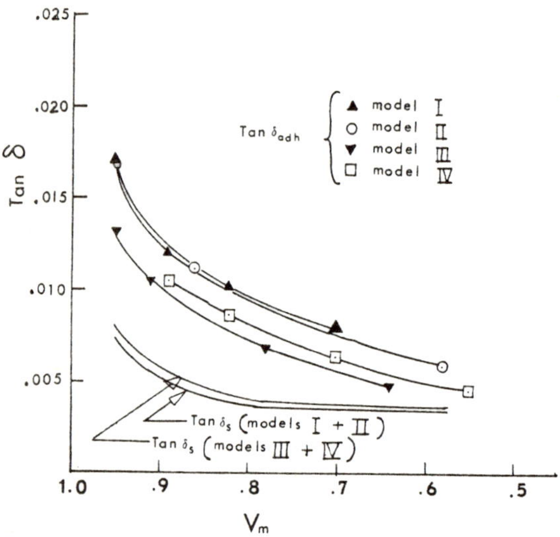

Fig. 4.13. Variation of tan δ_{adh} and tan δ_s with matrix volume fraction for copper wire composite model investigated.

1 to 3 times the internal energy dissipation due to the properties of the constituent materials themselves as represented by the parameter tan δ. These large differences further substantiate the appropriateness and feasibility of using the nondestructive-testing capability of instruments like the Rheovibron as a means of characterizing the degree of interface bonding in a filament-reinforced composite.

5. COMPOSITE STRUCTURE

Composite materials consist of two or more components and phases. The structure of a composite depends on the structure and arrangement of these components. Filled composites have a continuous matrix phase and a discontinuous filler phase made up of discrete particles. These include polyblends, filled polyvinyl chloride formulations (used in such application as floor tile and wire coating), filled rubbers, and glass-fiber-filled plastics [27].

Many composites contain fibers as the filler phase—for example, glass fibers in polyesters and epoxies, glass fabric laminates, asbestos fibers in phenolic resins, and chopped glass fibers in polystyrene and nylons. Recent developments are boron and graphite fibers in epoxy resins or polyimides, and whiskers (alumina or silicon carbide) in resins for aerospace applications [28]. The importance of fiber-filled composites arises largely from their unusually high strength and stiffness (modulus) per unit weight [29].

The dynamic mechanical properties of composite materials are determined by the properties of the components, the morphology of the system, and the nature of the interface between the phases. Therefore, a great variety of properties can be obtained by varying the structure of the system or the interface properties. As discussed in the previous sections, an important property of the interface is the degree of the adhesive bonding between the phases.

The effect of rigid fillers on dynamic mechanical properties of the filled rubber has been studied by Schwarzl et al. [30,31]. The results indicate that the dynamic shear modulus G at 1 Hz for a filled polyurethane rubber increases with increasing amounts of sodium chloride as filler. Fillers have a larger effect on the modulus above T_g than below it. The most pronounced effect of rigid fillers on the damping G"/G' is a broadening of the transition region by high concentration of fillers.

Fillers also decrease the damping as expressed by G"/G'. In this case the damping can be approximated by [32,33].

$$G''/G' = (G''/G')_1\phi_1 + (G''/G')_2\phi_2, \qquad (4.39)$$

where ϕ_1 and ϕ_2 are the volume fractions of the filler and matrix, and $(G''/G')_1$ and $(G''/G')_2$ are the loss tangents of the filler and matrix, respectively. However, Nielsen shows instances where fillers increase the damping of a mica-filled phenoxy resin [29]. The increased damping may result from particle-polymer friction. There are cases in which the filler shifts the loss-tangent peak and T_g to higher temperatures [32-35]. The shift in T_g is related to the surface area of filler. The effect should increase with concentration and the size of the particles.

Polyblends and block and graft polymers have two polymeric phases. These are composite structures. The dynamic mechanical properties of these polymers show two glass transitions, characteristic of the material in each phase, as seen in Fig. 3.21. The loss-tangent curves show two peaks [36-38]. The relative height of the two loss peaks is determined by the concentration of the two components and by the structure of the system.

Many of the fiber-filled composites are anisotropic (uniaxially oriented), with tremendous strength and modulus in one direction due to alignment of the reinforcing fibers. These composite structures are varied by a number of fabrication techniques [39-42]. Examples are laminates of sheets and woven fabrics, injection-molded polymers containing short fibers, and impregnated mats of fibers.

The dynamic modulus and loss factors in fiber-reinforced composites are influenced by the fiber orientations. The dynamic modulus and damping have been determined [43] by studying the free and forced transverse vibrations of cantilever beams made of the material, in which the fibers were uniformly oriented in the plane normal to the direction of vibration, making angles with the beam longitudinal axis of 0°, 22.5°, 45°, and 90°. The results are summarized in Table 4.9. The dynamic moduli decrease with decreasing fiber angle,

TABLE 4.9

Dynamic properties of fiber-reinforced composites at different fiber orientations

Beam material	Fiber orientation	Dynamic modulus E' (10^{11} dyn/cm^2)	Damping (10^{-2})
Aluminum		7.45	0.08
Epoxy		0.29	1.17
Composite	0°	3.55	0.16
	22.5°	1.91	0.65
	45°	1.16	0.92
	90°	1.02	0.85

while the damping factors increase. The effects of fiber orientation on the dynamic properties are determined by ultrasonic techniques [44-47].

A number of equations have been proposed for estimating static or dynamic moduli (48-50]. The longitudinal modulus E_L in tension on the uniaxially oriented system is given by

$$E_L = E_1\phi_1 + E_2\phi_2, \qquad (4.40)$$

where E_1 and E_2 are the moduli of matrix and fibers, respectively, and the corresponding volume fractions of the two phases are ϕ_1 and ϕ_2.

The transverse modulus E_T has been estimated by Nielsen [51,52]:

$$E_T E_1 = \frac{1 + AB\phi_2}{1 - B\psi\phi_2}, \qquad (4.41)$$

where

$$A = 0.5, \qquad (4.42)$$

$$B = \frac{E_2/E_1 - 1}{E_2/E_1 + A},$$

and

$$\psi = 1 + \left[\frac{1 - \phi_m}{\phi_m^2}\right]\phi_2, \qquad (4.43)$$

where ϕ_m is the maximum packing fraction of the fibers ($\phi_m = 0.785 \approx 0.907$).

The damping factor (tan δ) of the composite can be estimated by equation (4.21). Structural damping is expressed [53] as a structural loss coefficient η as follows:

$$\eta = \frac{D}{2\pi W}, \qquad (4.44)$$

where D is the energy dissipated in the composite system per cycle of vibration and W is the energy of vibration of the system.

The different composite structures are designed to improve various properties of materials: stiffness, strength, dimensional stability, impact strength and mechanical damping, etc. Tires, for example, are fiber-reinforced cord-rubber composites. They have three basic designs: bias-ply, radial-ply, and bias-belted. These structural designs affect the dynamic behavior of tire composites [54]. The dynamic moduli of rubbers and tire cords (nylon, polyester, rayon, and steel) influence the stiffness of a pneumatic tire [54-56], whereas the loss modulus and tan δ of these components are closely related to the amount of heat generated in the tire [57-59], as expressed by equation (1.19). High damping in a tire gives rise to an increase in its operating temperature; this can cause rapid degradation of the rubber and premature failure.

6. STRUCTURAL DAMPING AND ACOUSTICS

Polymeric materials are used as noise absorbers or damping materials in parts of automobiles, aircraft, railway equipment, air conditioning units, home appliances, etc. Noise reduction and the reduction of vibration resonances in these modern products are of environmental and industrial importance [60-63].

Noise is radiated by the vibration of an object. For example, if a clamped steel panel is subjected to vibration at one of its natural frequencies, the energy transmitted to the panel is stored as kinetic and potential energy. When the energy is supplied continuously, the vibration amplitude increases until the frictional losses equal the energy supplied. Internal friction (tan δ) in the steel being very small (0.001-0.004), most of the energy supplied is radiated into the air in the form of acoustic energy, which results in the generation of a loud sound even for very minute inputs of mechanical energy to the panel. However, if the panel is made of, or coated with a polymeric material presenting high internal friction (0.1-1.2), most of the energy will be lost inside the material instead of being radiated as acoustic energy, and the vibration amplitude will be much lower.

Since most metals have low internal losses, antivibration treatment, by the application of coating of damping materials called "deadeners", has received a great deal of attention in all metallic construction industries. Polymers are good damping materials. They absorb energy and then dissipate it as heat, resulting in decay of the amplitude of vibration. The greater the ability of a polymer to convert the energy into heat, the faster will be the decay.

Mastic deadeners, the most widely used of all materials made solely for the purpose of vibration damping, are supplied in semi-fluid form for application with a spray gun or trowel. Many mastic deadeners are of asphalt base with various fillers and fibers. The damping effectiveness of all asphalt-base mastic deadeners depends on the temperature. The temperature variations of the decay rate of two typical asphalt-base deadeners are shown in Fig. 4.14. A number of the viscoelastic foams (polyurethane-based foams) are excellent acoustical materials which can be used both as sound absorbers and for damping structural vibrations at the same time, due to their high internal friction and the interface friction between the air and cell walls [64,65]. These foams also provide the good thermal insulation and the vapor or liquid barrier afforded by closed-cell, self-skinning, hydrophobic polymeric materials.

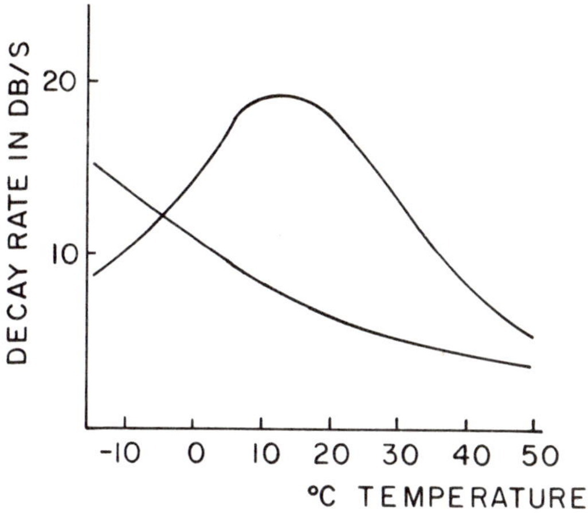

Fig. 4.14. Decay rate of two typical asphalt-base deadeners.

Acoustical properties of the polymer systems are related to the glass-transition temperature. Various polymer systems have been studied from this point of view by Chen and Williams [65]. They concluded that polymer systems should be designed to operate in the glass-transition region, as maximal energy dissipation takes place there. Thus, in designing damping systems to solve noise problems, one should first select homopolymers with proper T_g values and ranges, and then proceed by either polyblending the polymers, preparing copolymers, adding plasticizers, or adding fillers to achieve a broader T_g range for the expected use conditions. Numerous polymer systems with various acoustical properties are discussed in these articles [64,65]. The effect of internal friction (damping) in reducing noise and vibrations can be demonstrated by dropping on a hard floor a bar or other plastic object made from two different polymers—a low-damping polymer (polystyrene) and a high-damping polymer (cellulose nitrate). The low-damping polymer will give out a sharp ringing sound; the high-damping polymer will only produce a dull thud. These acoustical properties can be determined quantitatively by using the Bruel & Kjaer audio-frequency spectrometer. This unit measures the sound level in decibels (dB) and records sound spectrogram automatically from 20 Hz to 45,000 Hz [66].

There have been theoretical investigations concerning the sound absorption mechanism of materials [67], and information is available about the sound absorption characteristics of commercial products [68].

Aso and Kinoshita investigated the sound absorption characteristics of fiber assemblies [69,70]. The factors controlling these characteristics are the length of the fibers, their size, and their surface areas, as well as the porosity of the fiber assembly (quantity of fiber, bulk density), the state of arrangement of the fibers, the thickness of the assembly, and the frequency. For example, with high porosities the finer fibers show a higher rate of sound absorption because their total surface area is larger at a given porosity, and since the number density of fibers is high, the air spaces are finely divided. There is friction with the fiber surfaces due to the viscosity of the air, and the sound energy is thereby converted into heat energy as the main sound absorption process.

Plastic sheets, nonwoven fabrics, and papers generate noise during deformation. For example, if a piece of paper is subjected to a certain cyclic shearing deformation, the sound generated can be analyzed using an acoustic spectrometer [66]. Figure 4.15 shows

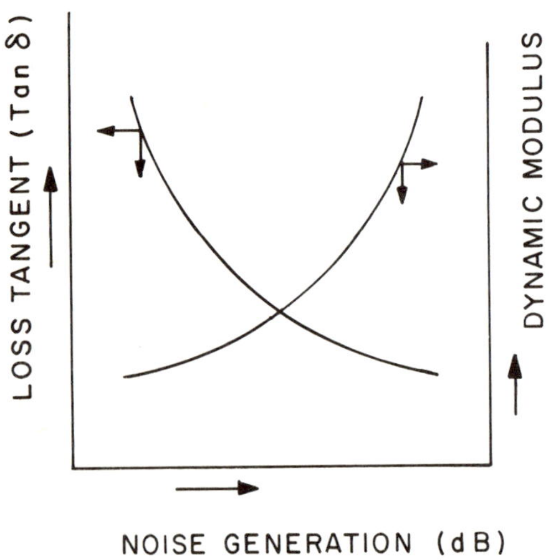

Fig. 4.15. Schematically the relationships between noise level and dynamic properties.

schematically the relationships between noise level and dynamic properties on these materials. Polymeric systems having higher stiffness and lower internal friction will generate higher sound pressure under a given deformation. These dynamic parameters of the systems are partly due to component properties and partly to assembly factors.

REFERENCES

1 J. E. Ruzicka, Structural Damping, Am. Soc. of Mech. Eng., New York, 1959.
2 A.S.T.M. Special Technical Publication No. 378, Internal Friction, Damping and Cyclic Plasticity, 1964.
3 J. W. Ballow and J. S. Silverman, Textile Res. J., 14 (1944) 282.
4 M. M. Platt, Textile Res. J., 20 (1950) 665.
5 M. Chaikin and N. H. Chamberlain, J. Textile Inst., 46 (1955) T25.
6 C. F. Zorowski and T. Murayama, Textile Res. J., 37 (1967) 852.
7 C. F. Zorowski and T. Murayama, Dynamics of Structured Solids (G. Herrmann, Ed.), Am. Soc. of Mech. Eng., New York, 1969, pp. 47-61.
8 J. S. W. Hearles, J. Textile Inst., 49 (1958) T389.
9 L. R. G. Treloar and G. Riding, Shirley Inst. Mem., 25 (1962).
10 W. J. Hamburger, Textile Res. J., 18 (1948) 705.
11 C. F. Zorowski, T. Murayama, and A. T. Alptekin, in Proc. 5th Int. Congr. Rheol., Vol. 3 (S. Onogi, Ed.), Univ. of Tokyo Press, Tokyo, 1970, p. 295.
12 J. L. White, C. C. Cheng, and K. E. Duckett, Textile Res. J., 46 (1976) 496.
13 C. F. Zorowski and T. Murayama, Dynamic Non-Destructive Bond Strength Characterization in Non-Woven Fabrics, paper presented at Fiber Society Meeting, Williamsburg, May 1974.
14 C. F. Zorowski and T. Murayama, Proc. 1st Int. Conf. on Mech. Behav. of Mater., Vol. 5, Soc. of Mater. Sci., Kyoto, Japan, 1972, p. 28.
15 A.S.T.M. D2138-67.
16 E. L. Brownlee, Modified Strip Adhesion Method, Presented to A.S.T.M. Subcommittee A-9, 17 October 1962.
17 T. Murayama and E. L. Lawton, J. Appl. Polym. Sci., 17 (1973) 669.
18 M. Takayanagi, Mem. Fac. Eng., Kyushu Univ., XIII, 1963.
19 T. Murayama, J. H. Dumbleton, and M. L. William, J. Macromol. Sci., B1 (1967) 1.
20 T. Takeyama and J. Matsui, Rubber Chem. Technol., 42 (1969) 229.
21 D. H. Kaelble, J. Macromol. Sci., C6 (1971) 85.
22 V. I. Feodosev, Selected Problems on Strength of Materials, Gostekhizdat, 1953.
23 D. N. Reshetov and Z. M. Levina, Damping of Vibrations in Joints of Machine Parts, Vestn. Mashinostr., 1 (1956) No. 12, p. 3.
24 G. Ya. Panovko, D. I. Goltzev, and G. N. Strakhov, Elementary Problems of Structural Hysteresis, in Voprosy Dinamiki i Prochnosti, V, Inst. of Machine Design, Acad. Sci. Latvian SSR, Riga, 1958, p. 5.
25 L. E. Goodman, A Review of Progress in Analysis of Interfacial Slip Damping, in Structural Damping, A.S.M.E., Dec. 1959, pp. 36-48.
26 A. T. Alptekin, On the Dynamic Mechanical Properties of Fiber Assemblies, Thesis, Mech. and Aerospace Eng. Dept., North Carolina State Univ., Raleigh, N.C., 1968.
27 L. Holliday (Ed.), Composite Materials, Elsevier, Amsterdam, 1966.
28 S. W. Tsai, J. C. Halpin, and N. J. Pagano, Composite Materials Workshop, Technomic, 1968.
29 L. E. Nielsen, Mechanical Properties of Polymers and Composites, Marcel Dekker, New York, 1974.
30 F. R. Schwarzl, H. W. Bree, C. J. Nederveen, G. A. Schwippert, L. C. E. Struik, and C. W. Van der Wal, Rheologica Acta, 5 (1966) 270.

31 F. R. Schwarzl, H. W. Bree, G. A. Schwippert, L. C. E. Struik, and C. W. Van der Wal, in Proc. 5th Int. Congr. Rheol., Vol. 3 (S. Onogi, Ed.), Univ. of Tokyo Press, Tokyo, 1970, p. 3.
32 L. E. Nielsen, R. A. Wall, and P. G. Richmond, SPE J., 11 (Sept. 1955) 22.
33 R. F. Landel and T. L. Smith, ARS J., 31 (1961).
34 G. Kraus and J. T. Gruver, J. Polym. Sci., A2, 8 (1970) 571.
35 S. A. Schreiner, P. I. Zubov, T. A. Volkova, and I. I. Vokulonskaya Colloid J. USSR (Engl. Transl.), No. 5, 26 (1965) 541.
36 T. Miyamoto, K. Kodama, and K. Shibayama, J. Polym. Sci., A2, 8 (1970) 2095.
37 R. Buchdahl and L. E. Nielsen, J. Appl. Phys., 21 (1950) 482.
38 E. Perry, J. Appl. Polym. Sci., 8 (1964) 2605.
39 L. J. Broutman and R. H. Krock, Modern Composite Materials, Addison-Wesley, Reading, MA, 1967.
40 P. Morgan, Glass Reinforced Plastics, Iliffe, London, 1961.
41 S. S. Oleesky and J. G. Mohr, Handbook of Reinforced Plastics, Reinhold, New York, 1964.
42 D. J. Duffin, Laminated Plastics, Reinhold, New York, 1966.
43 A. B. Schultz and S. W. Tsai, J. Compos. Mater., No. 3, 2 (1968) 368.
44 E. Behrens, Textile Res. J., 38 (1968) 1075.
45 J. M. Whitney, J. Compos. Mater., 1 (1967) 188.
46 A. W. Schultz, Mater. Res. Stand., 7 (1967) 341.
47 J. M. Whitney and M. B. Riley, AIAA J., 4 (1966) 1537.
48 S. W. Tsai, Structural Behavior of Composite Materials, NASA Rep. CR-71, 1964.
49 J. J. Hermans, Proc. Koninkl. Nederl. Adad. Wet., Amsterdam, 70B (1967) 1.
50 J. C. Halpin, J. Compos. Mater., 3 (1969) 732.
51 L. E. Nielsen, J. Appl. Phys., 41 (1970) 4626.
52 T. B. Lewis and L. E. Nielsen, J. Appl. Polym. Sci., 14 (1970).
53 E. M. Kerwin, Jr., Internal Friction, Damping and Cyclic Plasticity, STP 378, ASTM, PA, 1964.
54 S. K. Clark (Ed.), Mechanics of Pneumatic Tires, N.B.S. Monogr. 122, 1971.
55 P. Kainradl and G. Kaufmann, Rubber Chem. Technol., 49 (1976) 823.
56 P. R. Willett, J. Appl. Polym. Sci., 19 (1975) 2005.
57 P. R. Willett, Rubber Chem. Technol., 46 (1973) 425.
58 G. E. R. Lamb, H. D. Weigmann, and B. C. Goswami, Rubber Chem. Technol., 46 (1973) 425.
59 G. E. R. Lamb, H. D. Weigmann, and B. C. Goswami, Rubber Chem. Technol., 49 (1976) 1145.
60 R. N. Hamme, Harris Handbook of Noise Control: Vibration Damping, McGraw-Hill, New York, 1957.
61 M. P. Blake, and W. S. Mitchell, Vibration and Acoustic Measuring Handbook, Spartan, New York, 1972.
62 H. Oberst, Acustica, 2, Leaflet 4, Part I (1952), AB181-194; 4, Part II (1954) 433.
63 H. Oberst and K. Frankenfeld, Acustica, 4 (1952) 181.
64 A. C. F. Chen and H. L. Williams, J. Appl. Polym. Sci., 20 (1976) 3387.
65 A. C. F. Chen and H. L. Williams, J. Appl. Polym. Sci., 20 (1976) 3403.
66 Bruel and Kjaer, Audio Frequency Spectrometer Type 2112, Instruction and Applications, Copenhagen, 1974.
67 C. Zwikker and C. W. Kosten, Sound Absorbing Materials, Elsevier, 1949.
68 E. G. Richardson, Technical Aspects of Sound, Elsevier, 1953.
69 S. Aso et al., J. Textile Mach. Soc. Jap., 16 (1963) 861.
70 S. Aso et al., J. Textile Mach. Soc. Jap., 14 (1961) 285.

CHAPTER V

DYNAMIC PROPERTIES AND VISCOELASTICITY

1. INTRODUCTION

Polymer solids are deformed by external forces. Part of the work done is dissipated as heat. The solids also can store as well as dissipate mechanical energy. When polymers are subjected to sinusoidally oscillating stress, the strain is neither exactly in phase with the stress nor 90° out of phase, but somewhere in between. Some of the energy input is stored and recovered in each cycle, and some is dissipated as heat. Such behavior is known as viscoelasticity. Dynamic deformations and their viscoelastic responses are discussed in Chap. I.

The viscoelasticity of polymers is often studied by stress relaxation or creep. The relations between stress, strain, and their time dependencies are described by rheological equations. Extensive studies on this subject have been carried out by Ferry et al. Theoretical and experimental descriptions of viscoelasticity can be found in his book [1]. Mathematical studies of viscoelasticity have been made by Christensen [3], and several excellent books on viscoelasticity in polymers have been published [1-5]. An example of experimental relationships between dynamic properties and viscoelasticity is discussed in this chapter.

2. APPLICABILITY OF PHENOMENOLOGICAL THEORY

Most of the first few percent of deformation in nonpolymeric crystalline solids is nonrecoverable and little affected by changes in the time scale. However, the first few percent of deformation in polymeric amorphous solids is usually recoverable and quite dependent on changes in the time scale. This viscoelastic behavior of amorphous polymers is manifested in a variety of ways, such as (i) stress relaxation—the time-dependent decrease in the stress required to maintain a sample at constant deformation, (ii) creep—the time-dependent increase in deformation of a sample held at constant stress, or (iii) dynamic mechanical properties—the presence of both in-phase and out-of-phase components of the response to an oscillating stress or strain.

All of these experimental measurements can be interrelated using the phenomenological theory of linear viscoelasticity. This theory requires that stress and strain be linearly related at any given instant of time and that Boltzmann's superposition principle apply, i.e., that the total strain of a sample be an additive function of the individual strains resulting from each of a sequence of stress applications. The phenomenological theory states that the viscoelastic behavior of a real system can be represented mathematically by an array of Maxwell or Voigt elements each composed of an idealized Hookean spring with modulus G_i and a Newtonian dashpot with viscosity η_i. Each Maxwell or Voigt element is characterized by a relaxation or retardation time defined as $\tau_i = \eta_i/G_i$.

Empirically it has been found that an increase in temperature has the same effect on the measured viscoelastic behavior as an increase in time or decrease in frequency. This time-temperature superposition can be expressed in terms of the phenomenological theory by stating that increasing the temperature from T_o to T divides all relaxation or retardation times by the same factor a_T. The time-temperature superposition principle is discussed in Chap. I. Equation (1.24) gives the shift factor a_T.

The phenomenological theory of linear viscoelasticity has been applied to numerous amorphous polymers. Time-temperature superposition was found in many cases [1]. The deformation of oriented crystalline polymers is usually recoverable for moderate strains of a few percent; however, the dependence on change in time scale is much less than for amorphous polymers above T_g. Several investigations were undertaken to determine whether the phenomenological theory of linear viscoelasticity is applicable to oriented crystalline polymers [8-12]. Time-temperature superposition is useful for theoretical studies, and also for the prediction of the long-time properties (the lifetime) of polymers. In the following section we shall test the applicability of this theory to oriented nylon 66 fibers and oriented poly(ethylene terephthalate) fibers.

3. STRESS RELAXATION, CREEP, AND DYNAMIC PROPERTIES

(1) Nylon 66. The material used is a superbright (i.e., no TiO_2) nylon 66 yarn (70 total denier, 34 filaments) whose filaments have a birefringence of 0.058. It is thermally and mechanically condition[ed] by a modification of the procedure of Howard and Williams [7]. First the yarn is heated under no load in vacuum for 30 minutes at 150°C. Ten 50-cm lengths of yarn are plied together to make a sample. Such

samples are placed at five different positions in the stress-relaxation apparatus. The chamber is blanketed in dry nitrogen, and the temperature is raised to 150°C for a 30-minute period of equilibration. Then application of a 3.5×10^8 dyn/cm^2 stress produces creep, which continues for 2 hours. The stress is removed and creep recovery followed for 2 hours. The sample is cooled to room temperature and allowed to recover for another day.

Then stress relaxation measurements are made at 2% strain for 3 hours at 25°C. The first reading of stress is taken 4 seconds after strain application. Following stress relaxation, the strain is removed and the sample allowed to recover for 3 hours before heating to the next higher temperature. In this way, stress relaxation at 14 different temperatures in the range from 25 to 145°C for times of 1 to 3 hours are obtained, with the results shown in Fig. 5.1. The points represent an average of the five samples.

Inspection of these data indicates that time-temperature superposition ought to be applicable. Accordingly, 25°C is taken as the standard temperature, and each of the other curves is translated parallel to the log time scale by an appropriate amoung log a_T. No vertical shift was necessary. The resulting superposed curve is illustrated in Fig. 5.2, while the dependence of log a_T on temperature is shown in Fig. 5.3 and Table 5.1. The log a_T values used by Howard and Williams [7,8] to superpose creep data are also shown in Fig. 5.3. Both sets of points lie on the same straight line.

A tensile stress relaxation modulus E(t) can be calculated from a tensile creep compliance D(t) by means of the following equation [1,9]:

$$E(t) = \frac{\sin m\pi}{m\pi D(t)}, \qquad (5.1)$$

where m is the slope of a double logarithmic plot of D(t) against time t. For a comparable region of reduced time scale, the value of m is always less than 0.1 from the data of Howard and Williams [7,8]. Therefore, $(\sin m\pi)/m\pi \simeq 1$, and the modulus E(t) is simply the reciprocal of D(t). The modulus calculated in this way is compared with the experimental values in Fig. 5.4 and Table 5.2. Agreement is excellent, especially in view of the scatter of points around both superposed curves.

Dynamic mechanical properties were measured at a constant frequency of 11 Hz using the Rheovibron. The sample used was a semidull (i.e., containing 0.3% TiO_2) nylon 66 yarn (70 total denier, 34 filaments)

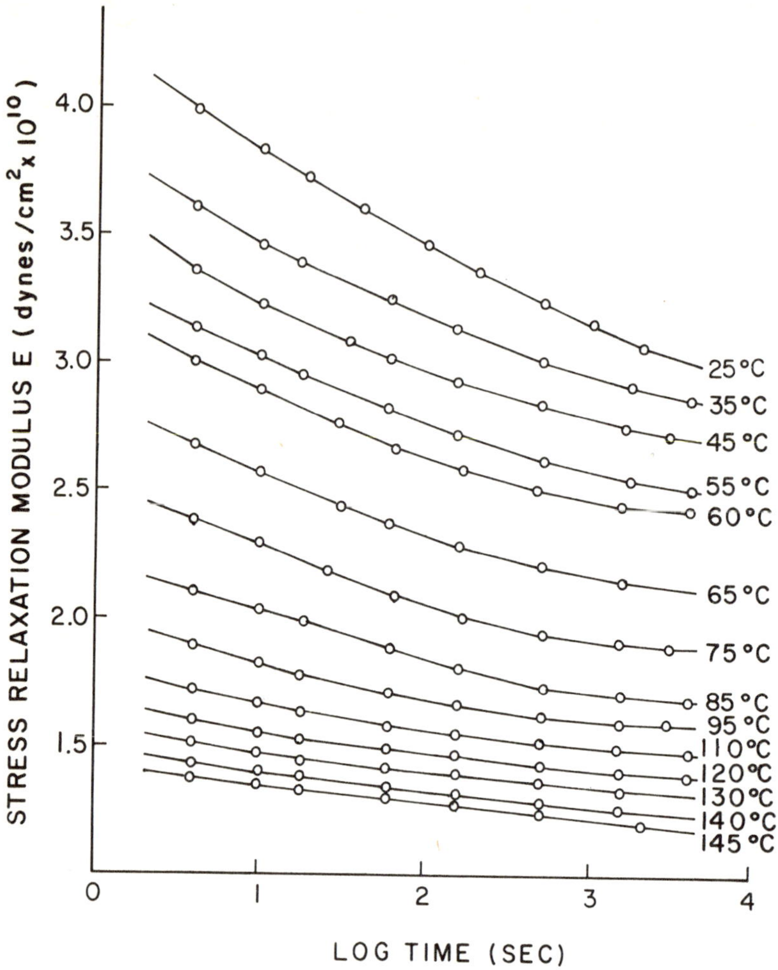

Fig. 5.1. Stress-relaxation curves for oriented nylon 66 at various temperatures.

whose filaments have a birefringence of 0.064. The sample was thermally conditioned by heating in a nitrogen atmosphere at 150°C under no load for 30 minutes. It was not mechanically conditioned, since the amount of strain in this dynamic experiment is quite small [10]. The dynamic modulus was measured in a nitrogen atmosphere at 5°C intervals from 25 to 140°C. The results are shown in Fig. 5.5.

To compare these data with those for stress relaxation and creep, it is necessary to convert the temperature dependence at constant frequency into a time (or frequency) dependence at constant temperature

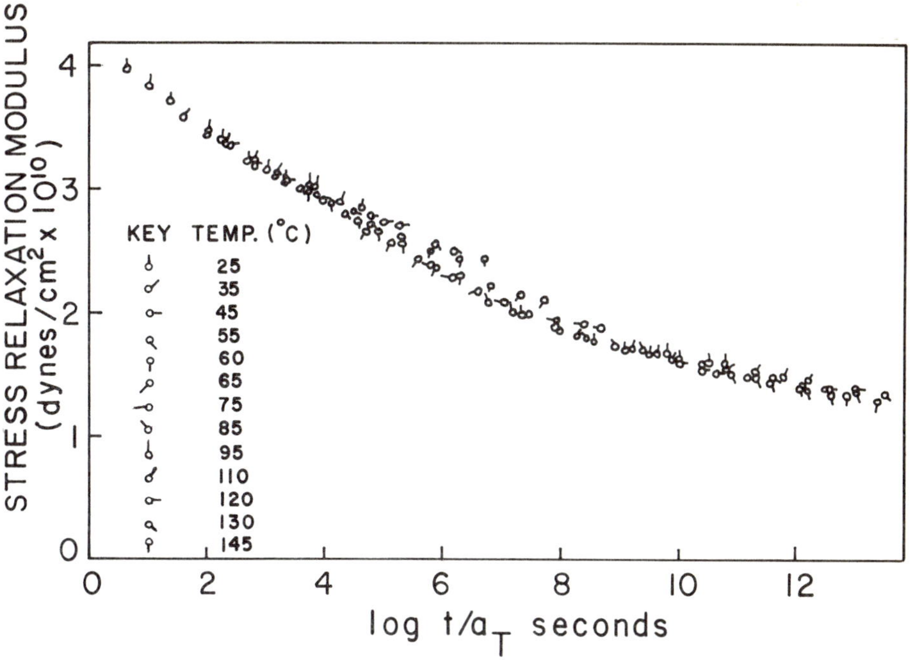

Fig. 5.2. Data of Fig. 5.1 superposed at 25°C using log a_T factors from Table 5.1.

The dynamic modulus at 25°C is fixed on the log time scale at -1.84, which corresponds to a circular frequency of 69 rad/sec. The increase in temperature from Fig. 5.5 is converted into a decrease in frequency or corresponding increase in time using the log a_T data of Table 5.1, which we know to be applicable.

Ferry [1] gives the following equation for converting the dynamic modulus $E'(\omega)$ to the stress-relaxation modulus $E(t)$:

$$E(t) = E'(\omega) - 0.40E''(0.40\omega) + 0.014E''(10\omega). \qquad (5.2)$$

However, for this nylon sample, E" is at least an order of magnitude smaller than E'; therefore, the last two terms of equation (5.2) are negligible. The stress-relaxation modulus calculated from this dynamic modulus is also shown in Fig. 5.4 and Table 5.2. Agreement with the experimental values from stress-relaxation and the values calculated from the creep compliance is quite satisfactory.

The phenomenological theory of linear viscoelasticity enables the time and temperature dependence of a viscoelastic function to be

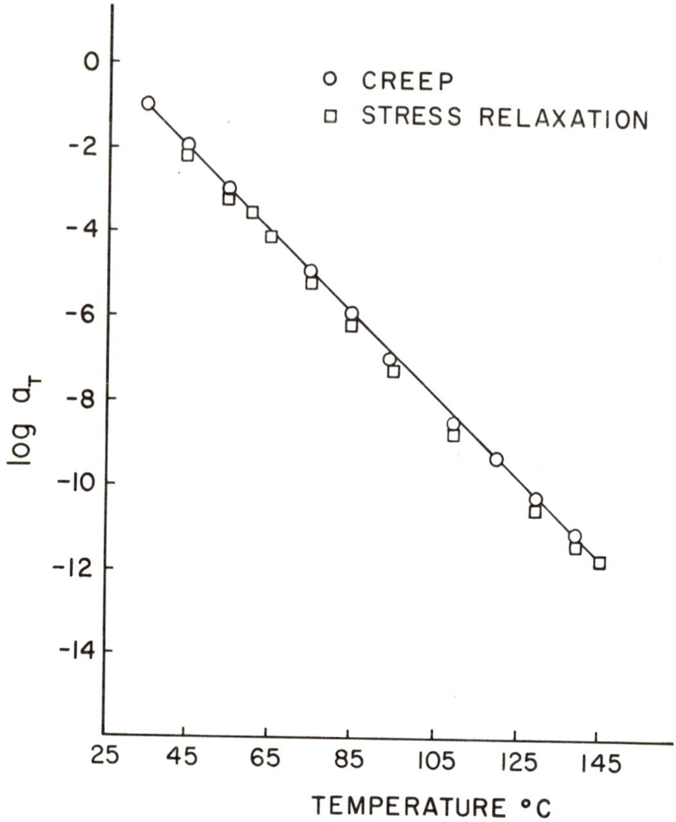

Fig. 5.3. Shift factor $\log_{10} a_T$ as a function of temperature.

TABLE 5.1

Values of log a_T used to superpose stress-relaxation data

Temperature (°C)	$\log_{10} a_T$	Temperature (°C)	$\log_{10} a_T$
25	0.0	85	-6.2
35	-1.0	95	-7.3
45	-1.8	110	-8.6
55	-2.6	120	-9.4
60	-3.1	130	-10.3
65	-4.1	140	-11.3
75	-5.2	145	-11.6

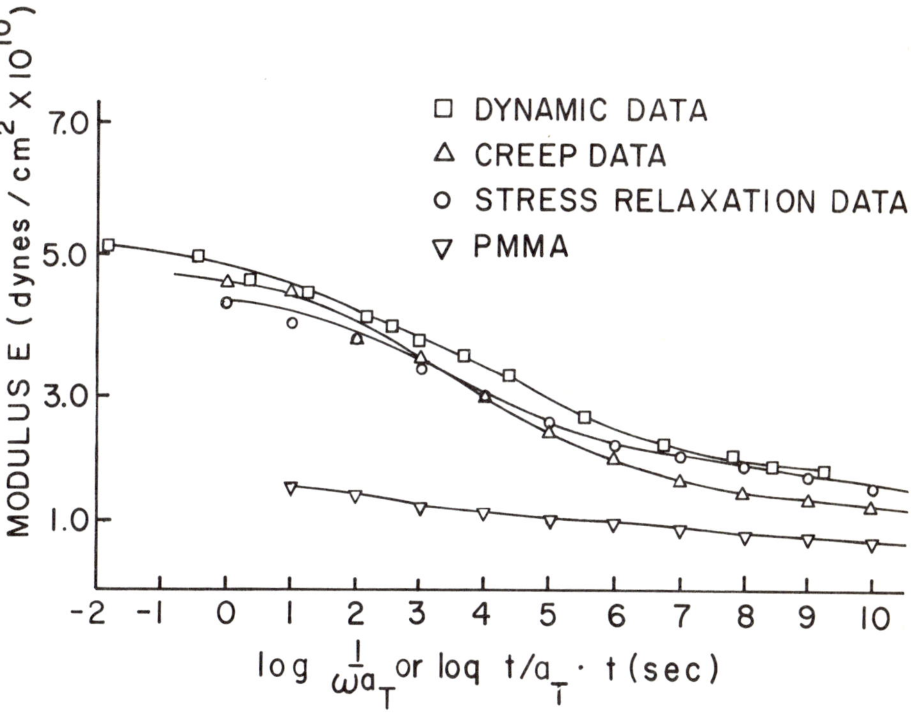

Fig. 5.4. Master curves at 25°C for oriented nylon 66.

separated. The result is the viscoelastic function (here the stress-relaxation modulus) over an extended time scale at some arbitrary reference temperature (here 25°C) plus the change in time scale, log a_T, of this curve when the temperature is changed. The major advantage of the phenomenological theory is the consequent interconvertibility of all experimental measurements of viscoelasticity.

The results illustrated here clearly indicate the applicability of this theory to oriented nylon 66 fibers. Time-temperature superposition is applied independently and successfully to stress-relaxation and creep data on different oriented nylon 66 fibers. The same viscoelastic function E(t) and the same values of log a_T are calculated from both sets of data. If these values of log a_T are used to superpose isochronal dynamic-modulus data, again a viscoelastic function E(t) can be calculated and is found to agree closely with the one previously obtained from stress-relaxation and creep. The three types of experiments were performed on different samples of

TABLE 5.2

Stress relaxation modulus calculated from creep and dynamic mechanical properties compared with experimental values

$\log_{10}$ [t(sec)]	E(t) (10^{10} dyn/cm^2)	1/D(t) (10^{10} dyn/cm^2)	E'(1/ω) (10^{10} dyn/cm^2)
-1.0		5.20	5.02
0.0		4.76	4.75
1.0	3.84	4.20	4.45
2.0	3.46	3.70	4.13
3.0	3.16	3.29	3.77
4.0	2.90	2.84	3.44
5.0	2.62	2.36	3.05
6.0	2.35	1.94	2.58
7.0	2.12	1.63	2.26
8.0	1.90	1.45	2.00
9.0	1.72	1.32	1.86
10.0	1.60	1.22	
11.0	1.50	1.14	
12.0	1.42	1.07	
13.0	1.35	1.01	
14.0	1.28	0.92	
15.0	1.20	0.83	

nylon 66 fibers, which had slight variations in orientation. However, all of them were subjected to essentially the same thermal-conditioning step. In addition, the samples for stress-relaxation and creep experiments were given the same mechanical conditioning. It is possible that the values of E(t) and log a_T may depend somewhat upon the degree of orientation and/or the details of the conditioning procedure. However, it is unlikely that the applicability of the phenomenological theory will be affected.

Comparison of the two functions E(t) and log a_T for nylon 66 fibers with the corresponding functions for typical amorphous polymers above T_g indicates striking differences. The dependence on time is much less for the nylon. The largest slope of a log E(t)-log t plot for nylon 66 is 0.08, whereas the slope is at least 0.5 for an amorphous polymer. The modulus of nylon 66 ranges from 1×10^{10} to 5×10^{10} dyn/cm^2 over this entire range of time scale. The modulus of the amorphous polymer approaches these values only at or below the transition temperature T_g. The temperature dependence of log a_T for nylon is linear, while it has the nonlinear WLF form for the amorphous polymer. Other than the applicability of the phenomenological theory, there are no similarities between nylon 66 and a typical amorphous polymer above T_g.

However, there are many similarities between this behavior of nylon 66 and the viscoelasticity of an amorphous polymer, such as

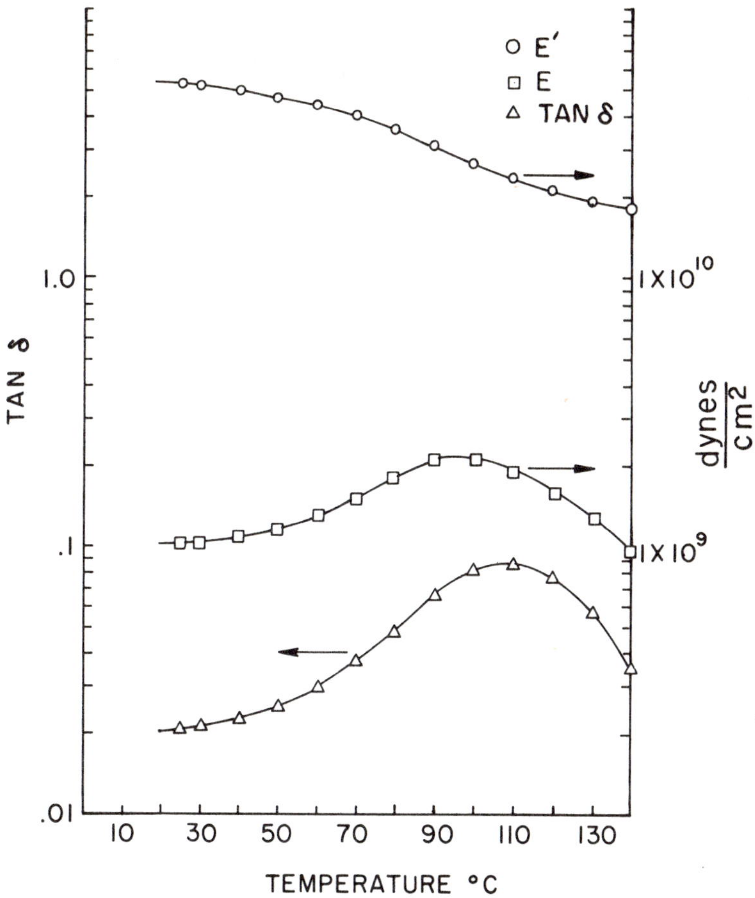

Fig. 5.5. Dynamic mechanical properties of oriented nylon 66 as a function of temperature.

poly(methyl methacrylate), below its T_g. Iwayanagi et al. [11-13] have measured both creep and stress-relaxation for this latter polymer over the temperature range from -51°C to +75°C. Time-temperature superposition could be applied independently to both sets of data. The viscoelastic functions E(t) from stress-relaxation and creep agreed, as is illustrated in Fig. 5.4 at a reference temperature of -51°C. The dependence on time is even less than for nylon 66. The modulus is also somewhat lower, although it is of the same order of magnitude of 10^{10} dyn/cm^2.

The values of log a_T used tc superpose the two sets of data also agreed. Furthermore, log a_T is found to be a linear function of temperature with a slope of -0.090, similar to the slope of -0.096 for nylon 66.

Of course, this almost quantitative agreement between oriented crystalline nylon 66 at 25°C and unoriented amorphous poly(methyl methacrylate) at -51°C may be fortuitous. However, similarities in macroscopic behavior between crystalline and glassy amorphous polymer are not new. Ender and Andrews [14] have observed equally suggestive agreement in plastic behavior.

The details of the molecular motion responsible for the viscoelasticity in both kinds of materials are still unclear. Evidence [15] clearly shows that crystal defects are responsible for both the α and γ loss mechanisms in melt-crystallized as well as single-crystal mats of polyethylene. For the γ-peak, the defects are dislocations within the lamellae. For the α-peak, the defects are located in the fold plane separating the lamellae. One important feature of all these defects is that they involve a small number of segments of a given polymer chain, although for a disclocation there must be a three-dimensional region of misfit centered on the dislocation line. However, dislocation lines can be postulated to move by the jumping of line segments or kinks rather than by simultaneous translation of the entire line [16]. In such a case, only a few chain elements need respond at any one time to an applied stress.

But this idea of a small region of a polymer chain being responsib for molecular motion is exactly the mechanism usually invoked for glassy amorphous polymers. The similarity in viscoelastic behavior could then be associated with the size of the molecular segment takir part in the motion. This motion for a crystalline polymer is constrained to follow crystallographic paths, whereas the motion for a glassy amorphous polymer would be less constrained. The comparison of nylon 66 and poly(methyl methacrylate) reported here seems to indicate that this constraint for nylon may be relatively unimportant. Confirmation of these suggestions awaits a detailed comparison of the molecular geometry of specific polymer chains.

(2) Poly(ethylene terephthalate) (PET). The material used was a PET yarn (560 denier, 110 filaments) drawn over a hot pin at 90°C with a machine draw ratio of 5, and subsequently annealed for 6 hours at 200°C under no tension. The structure of the fiber was characterized by x-ray techniques. The mass degree of crystallinity is 48% as measured by the procedure of Dumbleton and Bowles [17]. The radial half-width of the ($\bar{1}05$) reflection is 1.68°. If all this broadening is due to crystal size and if the factor K in the Scherre

equation is unity, then a crystal size of 51 Å can be calculated [18].

The half-width of the azimuthal scan of the ($\bar{1}05$) reflection is taken as a measure of the crystallite orientation. The intensity distribution is Gaussian to a good approximation. The half-width can be used to characterize the Gaussian distribution and $\langle\cos^2 \phi\rangle$ obtained by numerical integration [19]. The chain axis makes an angle ϕ with the fiber axis. The crystalline orientation factor f_c, defined as $\frac{1}{2}(3\langle\cos^2\phi\rangle - 1)$, is 0.95 for this sample. For complete orientation, f_c would be unity. The annealing step did not affect the crystalline orientation.

The total birefringence is the sum of contributions from crystalline and amorphous regions [20]:

$$\Delta n = \beta f_c \, \Delta n_c + (1 - \beta) f_a \, \Delta n_a . \qquad (5.3)$$

Here β is the volume percent crystallinity, f_a is the unknown amorphous orientation factor, and Δn_c and Δn_a represent the birefringence of a completely oriented sample which is completely crystalline or amorphous respectively. These latter factors are estimated to be 0.200 and 0.275 respectively [21], from the method of Samuels [22]. The fact that Δn_a is greater than Δn_c may be ascribed to internal-field effects as discussed by Stein [23]. The measured birefringence of the yarn is 0.193, and f_a is calculated as 0.71. From the experimental values of f_c and f_a it is concluded that this sample is highly oriented.

The dynamic mechanical properties—namely, the loss tangent tan δ and modulus E'—were measured at 0% RH using a Rheovibron. A temperature range of 22 to 165°C was scanned at each of four frequencies: 3.5, 11, 35, and 110 Hz.

Stress-relaxation was measured using the Vibron instrument and a stress-recording device illustrated in the block diagram of Fig. 5.6. The output voltage of the Vibron strain gauge is converted into a frequency, which is displayed on the counter. The calibration is such that the load in grams can be read directly from the counter. The digital printer, which records both stress and time, is operated by the counter by means of the digital clock.

The proportionality between output voltage and load was confirmed by calibration with known static loads. The stability of the system was tested by observing the variation of output with time for a known static load. From these measurements, the error in load measurement is less than 3%, and there is negligible drift of the load reading over a period of 24 hours.

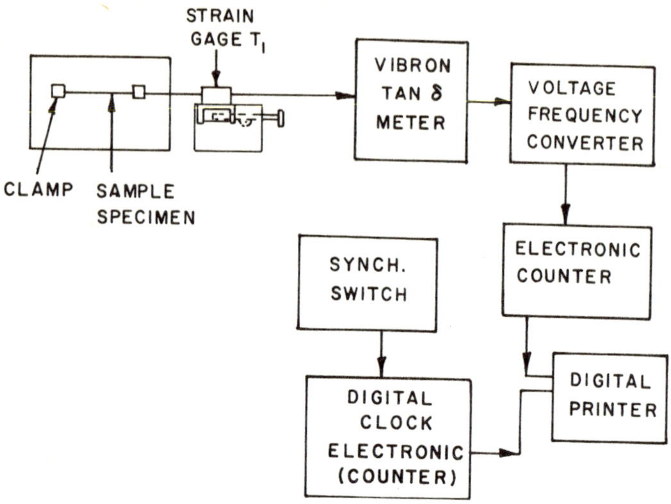

Fig. 5.6. Block diagram of stress relaxation apparatus.

A specimen about 5 cm long is mounted in the Rheovibron clamps. One clamp is attached to the strain gauge for load measurement. The other clamp is connected to a slide attached to a scale. This slide can be moved to impose a fixed extension of 1% as measured by the scale. It is fixed in position during the stress-relaxation measurements. The sample and both clamps are inside a chamber fitted with heaters capable of maintaining a constant temperature (±1°C) up to 200°C with an appropriate controlling device.

Measurements were begun at 22°C. The first reading of stress was taken 5 seconds after strain application. After following stress-relaxation for 3 hours, the strain was removed and the sample allowed to recover for 20 hours before heating to the next higher temperature. In this way, stress relaxation was obtained at 14 different temperatures from 22 to 182°C at 0% RH.

The results of the stress-relaxation experiments are shown in Fig. 5.7. The curve at 82°C is arbitrarily selected to remain fixed on the time scale; the curves at other temperatures are shifted parallel to the log time axis to obtain the superposed curve shown in Fig. 5.8. McCrum [24] has suggested that a vertical shift factor is necessary for superposition. However, empirically such a shift is not required for these data. The values of log a_T, the time-scale shift factor, are given in Table 5.3 and plotted against temperature in Fig. 5.9.

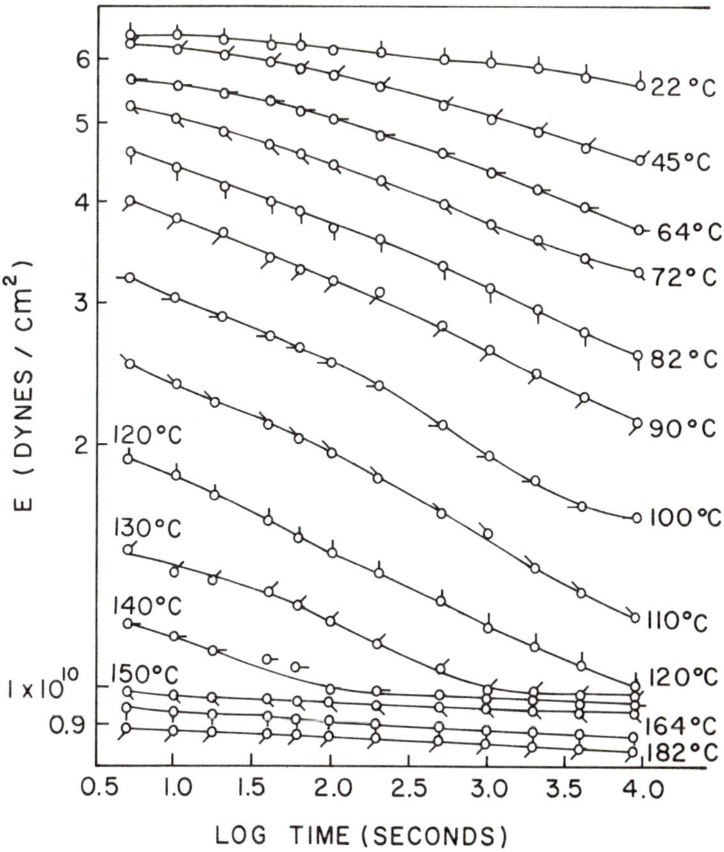

Fig. 5.7. Stress relaxation modulus as a function of log time at 14 temperatures as indicated.

The dynamic modulus E' and loss tangent tan δ are shown in Fig. 5.10. These isochronal curves as a function of temperature are transformed into isothermal curves as a function of time by applying the shift factors of Table 5.3. The result is a superposed curve represented by the points in Fig. 5.11. This procedure successfully accounts for the decade-and-a-half variation in frequency of the dynamic experiments. Furthermore, the curve of dynamic-modulus versus time is practically the same as that of stress-relaxation modulus versus time. This similarity is to be expected, since the difference between the two quantities depends only on E" (= E' tan δ), which is an order of magnitude smaller than E' for oriented PET.

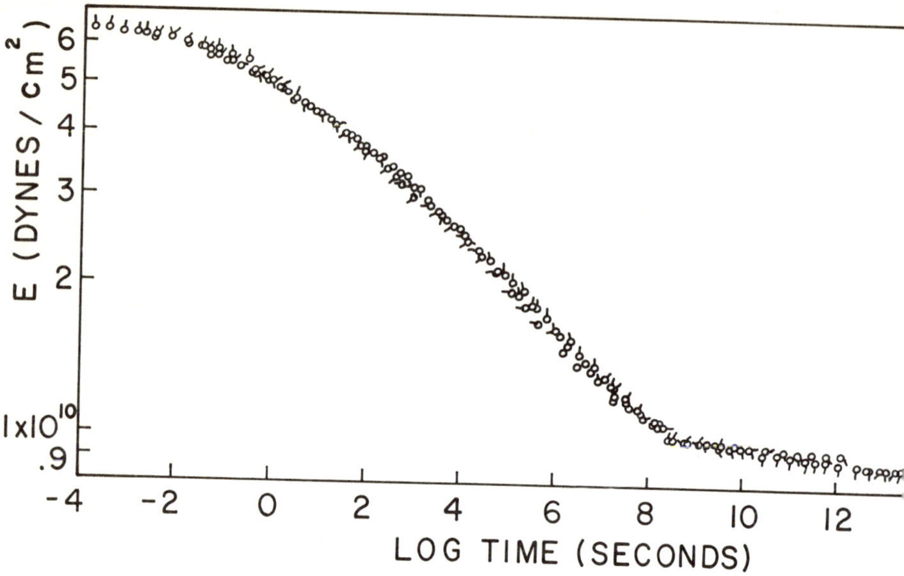

Fig. 5.8. Data of Fig. 5.7 superposed to give the time dependence of an isothermal modulus at 82°C. Pips as in Fig. 5.7.

TABLE 5.3

Values of log a_T used to superpose stress-relaxation data

Temp. (°C)	$\log_{10} a_T$	Temp. (°C)	$\log_{10} a_T$
22	4.48	110	-3.32
45	3.14	120	-4.52
64	2.00	130	-5.48
72	1.08	140	-6.54
82	0.00	150	-8.18
90	-0.84	164	-9.78
100	-2.06	182	-11.96

These properties of oriented PET fibers are quite similar to those of oriented nylon 66 previously studied [6]. For both, log a_T is found to be linear in temperature, with a slope of 0.116 for PET and 0.096 for nylon. There is an unexplained divergence from linear behavior for temperatures less than 60°C for the PET. The modulus-time curve for PET at 82°C is almost the same as that of nylon 66 at 25°C, as seen in Fig. 5.11. The PET has a slightly sharper dispersion region. This comparison clearly shows that the hydrogen bonds in nylon have only a slight effect on the shape of either the log a_T --temperature or the modulus-time curves. The major difference in

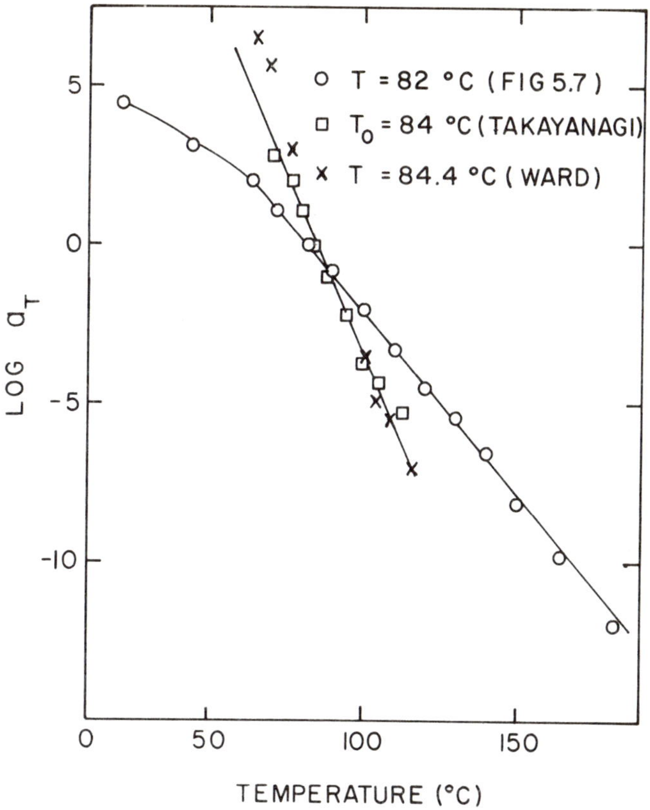

Fig. 5.9. Shift factor log a_T required to superpose viscoleastic data: (O) from Fig. 5.7; (□) stress relaxation from Takayanagi [10], (X) creep from Ward [25].

the two materials is the temperature required to fix the dispersion region at identical times.

There have been many studies of the viscoelasticity of both unoriented and oriented PET samples. However, the majority of these reported isochronal data insufficient to calculate the time dependence. The reports of Ward [25] and Takayanagi [10] are extensive enough for a comparison to be made with our data. Ward [25] measured the creep compliance of unoriented PET with an x-ray crystallinity of 33%. The sample had been heat-treated to 180°C for 1 hour. Takayanagi [10] measured the stress relaxation of unoriented PET samples with 5, 38, and 45% crystallinity as determined from the density. The samples with 38 and 45% crystallinity were obtained by heating for 24 hours at 100 and 130°C respectively.

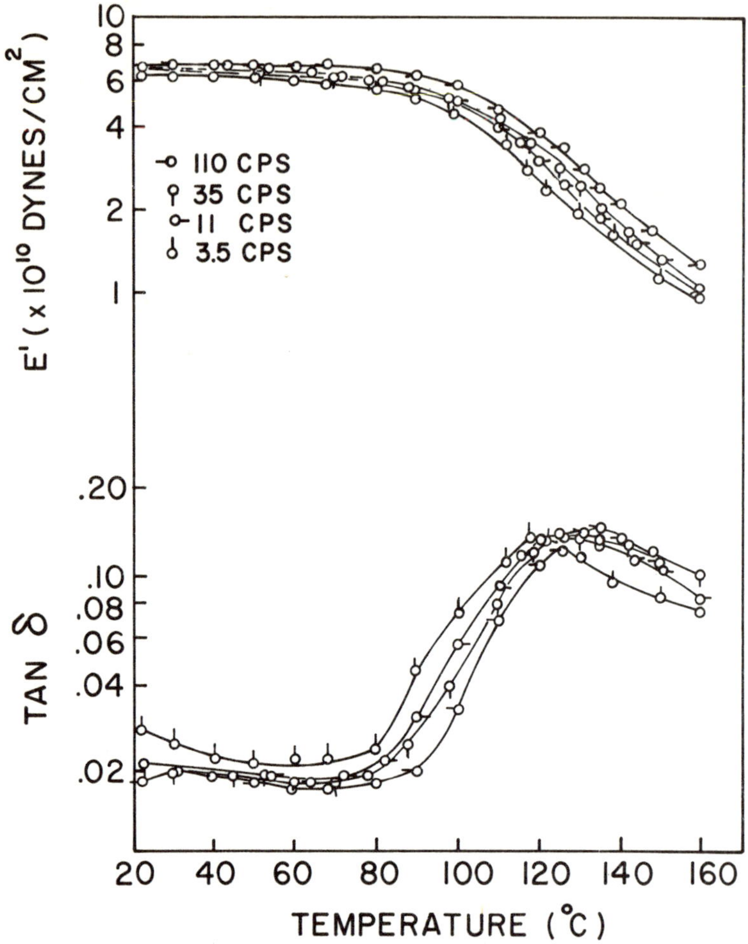

Fig. 5.10. Dynamic modulus E' and loss tangent tan δ as a function of temperature for four different frequencies as indicated.

Takayanagi [10] found that the dispersion region of the modulus curve was broadened considerably when the crystallinity increased from 5 to 38%. However, a further increase to 45% had only a slight effect. Therefore, we assume, as a first approximation, that the difference in crystallinity between our sample and those of Ward and Takayanagi had a minor effect on the viscoelastic properties. Any differences in these properties are thus attributed to the orientation.

Both Ward and Takayanagi indicate that the WLF equation fits their log a_T data. As illustrated in Fig. 5.9, a straight line would

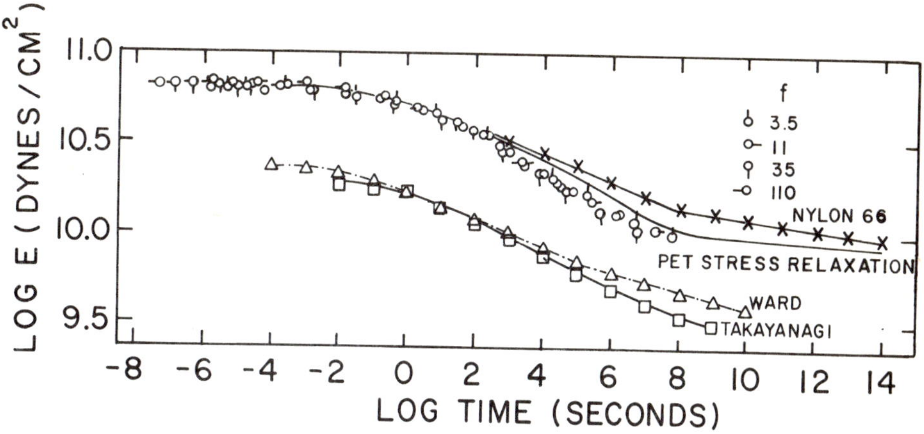

Fig. 5.11. Isothermal modulus as a function of log time for various samples (——) oriented PET at 82°C from Fig. 5.8; (Ǫ, ♁, O-, -O); oriented PET at 82°C from superposition of data of Fig. 5.10; (x) oriented nylon 66 at 25°C [7]; (△) unoriented PET at 84.4°C [25], (□) unoriented PET at 84°C [10].

do equally well; however, the slope would be 0.225, almost twice what was found for the oriented sample. The temperature range covered by both Ward and Takayanagi is so small that it is difficult to decide on the best relationship. Ward combined his log a_T data from creep measurements with the temperature dependence of the dielectric-loss peak obtained by Reddish [26] in order to extend the temperature range. Thereby, he definitely found the WLF equation to fit. Regardless of the specific form of the log a_T--temperature curve, it is clear that the viscoelasticity of the oriented sample is much less temperature-dependent.

The modulus is calculated from the reciprocal of the creep compliance measured by Ward, and in Fig. 5.11 it is compared with the stress-relaxation modulus of our oriented sample and the unoriented 45%-crystalline sample of Takayanagi. The two curves for the unoriented sample are in good agreement. The modulus of the oriented sample has exactly the same shape as that of Takayanagi's sample over the entire time scale; however, it is larger in magnitude by a factor of 3.3.

Based on this comparison, we conclude [27] that, at constant crystallinity, the effect of orientation is (1) to increase the magnitude of the modulus without changing its time dependence and (2) to decrease the temperature dependence of the modulus. From the isochronal measurements by Thompson and Woods [28], Newman and

Cox [29] conclude that there is a distinct interaction between crystallinity and orientation in PET. A quantitative measurement of any such interaction will require isothermal data over a wide time scale for samples of widely differing orientation and crystallinity.

4. STRESS-STRAIN BEHAVIOR

The studies of viscoelasticity in polymers have demonstrated that the stress-strain relations are determined as a function of time and temperature. This was discussed in the previous section. The magnitude of the initial strain in stress-relaxation and that of the load in creep are important factors for these experiments. The levels of applied stress on viscoelastic bodies in dynamic mechanical analysis yield a dynamic response, linear or nonlinear. The relationships between stress and strain in dynamic analysis are discussed in the section on stress-strain oscillation in Chap. I.

The stress-strain behavior in mechanical testing is commonly expressed by stress-strain curves. Types of these curves indicate whether materials are hard and brittle, ductile, or soft, as well as their modulus, elongation, and strength. Nielsen has given comprehensive reviews of stress-strain curves of many polymers under different conditions [30]. The combination of dynamic mechanical analysis and strain-stress tests is useful for obtaining the dynamic properties at different strain and stress levels [31].

The experiments are carried out by using a tensile testing machine with the PPM-5 (pulse propagation meter) Dynamic Modulus Tester. Figure 5.12 represented the mechanism of dynamic analysis of linear and nonlinear stress-strain characteristics for a polymeric solid. In this case, if the solid is subjected to a periodic strain variation of magnitude ε_0 imposed on a constant strain $\bar{\varepsilon}$ in the linear region of the curve, the stress and strain in the solid will oscilla between points A and B with the modulus E. If the same periodic strain variation is imposed on a solid subjected to a constant strai $\bar{\varepsilon}'$ in the nonlinear region, the following behavior is postulated: As the first strain variation propagates along the solid, the stress increases from point A' to B' and then relaxes along the path B' to A", much as in a crystalline solid. Subsequent ε_0 strain variation will produce stress-strain variations along B' to A" with the modulu $\bar{E}$. Thus, whether the solid is responding in its linear or its nonlinear region, it will propagate periodic strain variations as if it were a linear material with a modulus dependent on the imposed initial constant strain level. The dynamic moduli E and $\bar{E}$ can be

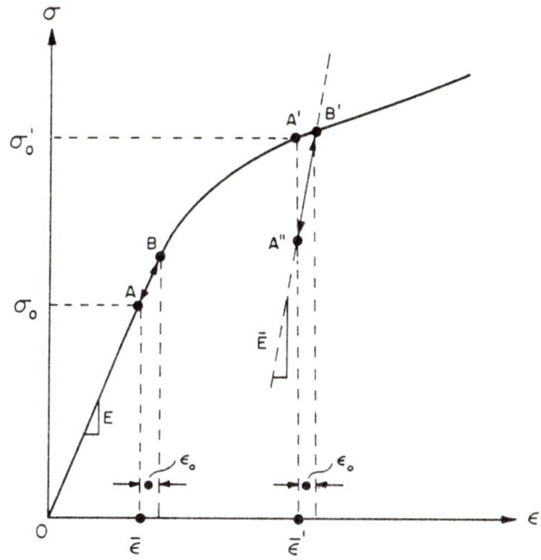

Fig. 5.12. Filament stress-strain variations resulting from periodic wave propagation.

measured directly using a high-frequency pulse-propagation technique. They have been shown to be dependent on initial filament strain. Figure 5.13 shows a typical variation of the dynamic modulus of

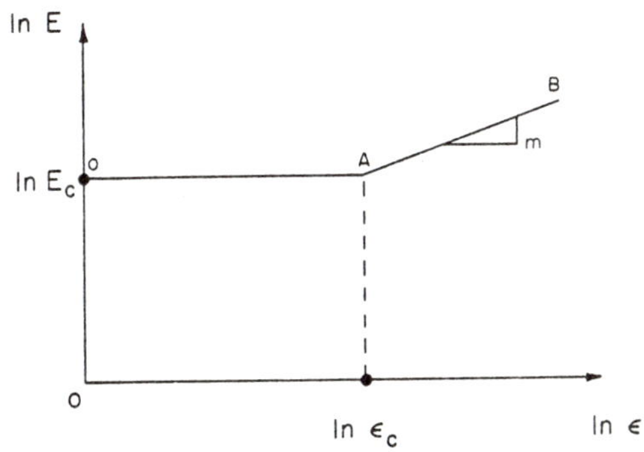

Fig. 5.13. Typical dynamic modulus variations with filament strain.

polymer solids with strain. The dynamic modulus is almost independent of strain from 0 to A. This corresponds to the linear elastic region of the material. When the strain exceeds ε_c, ln E increases linearly with ln ε.

As the strain increases, some polymers change in molecular arrangement. The molecules tend to become highly oriented parallel to the stretching direction. This orientation often gives high modulus and lower internal friction.

REFERENCES

1 J. D. Ferry, Viscoelastic Properties of Polymers, 2nd ed., Wiley, New York, 1970.
2 A. V. Tobolsky, Properties and Structure of Polymers, Wiley, New York, 1960.
3 R. M. Christensen, Theory of Viscoelasticity: An Introduction, Academic, New York, 1971.
4 J. J. Aklonis, W. J. MacKnight, and M. Shen, Introduction to Polymer Viscoelasticity, Wiley, New York, 1972.
5 H. Leaderman, Physics of High Polymers, Tokyo Inst. of Technol., Tokyo, 1957.
6 T. Murayama, J. H. Dumbleton, and M. L. William, J. Macromol. Sci. (Phys.), B1(1), (1967) 1-14.
7 W. H. Howard and M. L. Williams, Textile Res. J., 33 (1963) 689.
8 W. H. Howard and M. L. Williams, Textile Res. J., 36 (1966) 691.
9 J. J. Aklonis and A. V. Tobolsky, J. Appl. Phys., 37 (1966) 1949.
10 M. Takayanagi, Mem. Fac. Eng., Kyushu Univ., 23 (1963) 41.
11 S. Iwayanagi, J. Sci. Res. Inst. Jap., 49 (1955) 4.
12 K. Sato, H. Nakane, T. Hideshima, and S. Iwayanagi, J. Phys. Soc. Jap., 9 (1954) 413.
13 N. Saito, K. Okano, S. Iwayanagi, and T. Hideshima, Solid State Physics, Advances in Research and Applications, Vol. 14, Academic, New York, 1963.
14 D. H. Ender and R. D. Andrews, J. Appl. Phys., 36 (1965) 3057.
15 K. M. Sinnott, J. Appl. Phys., 37 (1966) 3385.
16 A. D. Brailsford, J. Appl. Phys., 36 (1965) 3941.
17 J. H. Dumbleton and B. B. Bowles, J. Polym. Sci., A4 (1966) 951.
18 H. P. Klug and L. E. Alexander, X-Ray Diffraction Procedures, Wiley, New York, 1954.
19 D. R. Buchanan, private communication.
20 G. R. Taylor and S. R. Darin, J. Appl. Phys., 26 (1955) 1075.
21 J. H. Dumbleton, J. Polym. Sci., Part A-2, 6 (1968) 795.
22 R. J. Samuels, J. Polym. Sci., A3 (1965) 1741.
23 R. S. Stein, J. Polym. Sci., C15 (1966) 185.
24 N. G. McCrum and E. L. Morris, Proc. R. Soc., 281A (1964) 258.
25 I. M. Ward, Polymer, 5 (1964) 59; P. R. Pinnock and I. M. Ward, Polymer, 7 (1966) 255.
26 W. Reddish, Trans. Faraday Soc., 46 (1950) 459.
27 T. Murayama, J. H. Dumbleton, and M. L. Williams, J. Polym. Sci., Part A-2, 6 (1968) 787.
28 A. B. Thompson and D. W. Woods, Trans. Faraday Soc., 52 (1956) 1383.
29 S. Newman and W. P. Cox, J. Polym. Sci., 46 (1960) 29.
30 L. E. Nielsen, Mechanical Properties of Polymers and Composites, Marcel Dekker, New York, 1974.
31 C. F. Zorowski and T. Murayama, in Dynamics of Structured Solids, (G. Herrmann, Ed.), Am. Soc. of Mech. Eng., New York, 1969, pp. 47-61.

CHAPTER VI

DYNAMIC ANISOTROPIC BEHAVIOR

1. INTRODUCTION

Numerous polymers are anisotropic, i.e., have different properties in different directions [1-5]. Examples include fibers, wood, oriented amorphous polymers, crystalline polymers, injection-molded plastics, fiber-filled composites, and single crystals. These materials have a minimum of five elastic moduli. The exact number of independent moduli depends upon the symmetry in the system [2,3]. Theoretically the elastic tensor can have as many as 36 independent components, but there is much greater symmetry than that in most practical engineering situations. The tensorial representation of elasticity may be in terms of either moduli or compliances [2].

The elastic constants for specimens possessing fiber symmetry or orthorhombic symmetry are discussed in detail by Ward [2,9]. The first attempts to determine a set of elastic constants for oriented polymers were Raumann's measurements on low-density polyethylene sheets [6,7] and the combination of measurements on fibers and films of polyethylene terephthalate undertaken by Pinnock and Ward [8,9]. Extensive studies on the anisotropic mechanical behavior of polymers were carried out by Ward and coworkers [10-12]. Ward also reviewed these properties in his book [2]. Dynamic mechanical analyses have been used to study anisotropic properties of polymers. Longitudinal-wave-propagation methods (sonic technique) are commonly applied to determine the influence of molecular orientation on the extensional moduli of fibers drawn to different draw ratios [13-19].

However, the effects of anisotropy on internal friction and loss modulus in low-frequency oscillation have been studied comparatively little [20,21]. In this chapter, therefore, recent developments of experimental techniques on dynamic shear, compression, and bending methods using low-frequency oscillation will be discussed, along with the experimental results.

2. TENSION

Several dynamic mechanical analyzers apply a sinusoidal tensile strain to one end of a sample and measure the stress output at the other end. The Rheovibron viscoelastometer is a typical example for obtaining dynamic tensile mechanical properties of films and fibers over a wide temperature range from -160°C to 250°C in an atmosphere at 0% relative humidity. The instrument operates at frequencies of 3.5, 11.0, 35.0, and 110 Hz. These tensile-mode instruments can be used to determine anisotropic behavior of polymers using film specimens cut in directions different from the stretch direction.

Takayanagi et al. [20] studied the dynamic mechanical properties of a series of drawn sheets and drawn and annealed sheets of high-density polyethylene, polypropylene, polyoxymethylene, poly(ethylene oxide), and polytetrahydrofuran. Uniaxial orientation was assumed in all cases. The dynamic extensional modulus was measured along the draw direction and perpendicular to it. It was found that for drawn and annealed samples, the dynamic tensile modulus was lower along the stretch direction than perpendicular to it for temperatures above the primary dispersion temperature. The parallel modulus E_0 crosses the perpendicular modulus E_{90} in the glass transition region. Therefore, although $E_0 > E_{90}$ at low temperatures (as expected from the molecular orientation), $E_0 < E_{90}$ at high temperatures. Takayanagi proposed a model to explain this behavior [20]. Schematically, in this model, the application of stress in the parallel direction results in a large fall in modulus with increasing temperature, the stiffness at high temperatures being primarily determined by the amorphous regions, which are very compliant above the relaxation transition. In the perpendicular direction the fall in modulus is not so great, as the crystalline region still supports the applied stress and a relatively high stiffness is maintained. This model correctly predicts the crossover of the dynamic moduli.

Gupta and Ward [22] observed similar crossover points in drawn and annealed sheets of low-density polyethylene and also in orthorhomic drawn, rolled, and annealed sheets of this polymer. They attributed the fall in modulus in the parallel direction to an interlamellar shear process. Structural studies [23,24] imply that these annealed sheets possess a lamellar texture, low-angle x-ray diffraction measurements showing that the planes of the lamellae make angles of approximately 40° with the initial draw direction. Thus when the tensile stress is applied along the initial draw direction, the direction of maximum shear stress is approximately parallel to the lamellar planes. At temperatures above the relaxation transition (which involves mobility of material between the lamellae) interlamellar shear can occur, with a corresponding fall in the modulus. Application of a tensile stress parallel to the initial draw direction results in a maximum shear stress nearly parallel to the lamellar planes. Application of the tensile stress in the 90° direction gives maximum shear stress parallel to only some lamellar planes. This explains the observation that $E_0 < E_{90}$ above the relaxation transition.

Stachurski and Ward [25,26] confirmed these conclusions by studying the viscoelastic behavior of these sheets. They reported that the anistropy of the loss factor tan δ gave a striking characterization for the assignment of the relaxation to different molecular processes. There are two relaxation processes in this region, with very different anistropy. These are attributed to the interlamellar shear relaxation and the C-axis shear relaxation respectively.

Dumbleton and Murayama [27] studied the effect of the anistropy on nylon 66 films and fibers. In this study samples are prepared as follows: Nylon 66 film 10 cm wide was drawn on an Instron testing machine at 100%/min, under standard conditions (22°C, 65% RH), using a 5-cm gauge length. Ink dots were placed at 5-mm intervals along the length of the specimen so that the draw ratio could be measured. Drawing at room temperature gave a maximum draw ratio of 3X, and it was necessary to heat the film during drawing to produce a higher draw ratio. Accordingly, samples were drawn in an oven on the Instron at 200°C. The highest draw ratio obtained was 5.5X, but subsequent relaxation of the sample, on removal from the clamp, reduced this to 4.5X. Sample nylon 66 yarns drawn, at room temperature, up to a draw ratio of 4.74X were also employed in studies to determine the effect of orientation on the dynamic behavior.

Dynamic mechanical measurements were made at 0% RH on a Rheovibron which measured the dynamic tensile modulus E' and the loss factor tan δ. The loss modulus E" was calculated from the relation E" = E' tan δ. Measurements were made at a fixed frequency of 11 Hz from room temperature at 10°C intervals, except near the transition region, where smaller intervals were employed.

The dynamic properties of the unoriented nylon 66 film displayed the α-transition peak, indicated by the maximum in E", at about 80°C. This is typical of the values obtained for similarly prepared nylon 66 [28-30]. Figure 6.1 shows the results for the cold-drawn 3X samples, cut parallel and perpendicular to the stretch direction. The E" curves are broader than for the unoriented film, and the maximum in E" occurs near 100°C for both samples. The E' curves do not cross and, in fact, run parallel to one another, with E' always higher in the direction of stretch. The Rheovibron data for the sample drawn 4.5X at 200°C are shown in Fig. 6.2. The maximum in

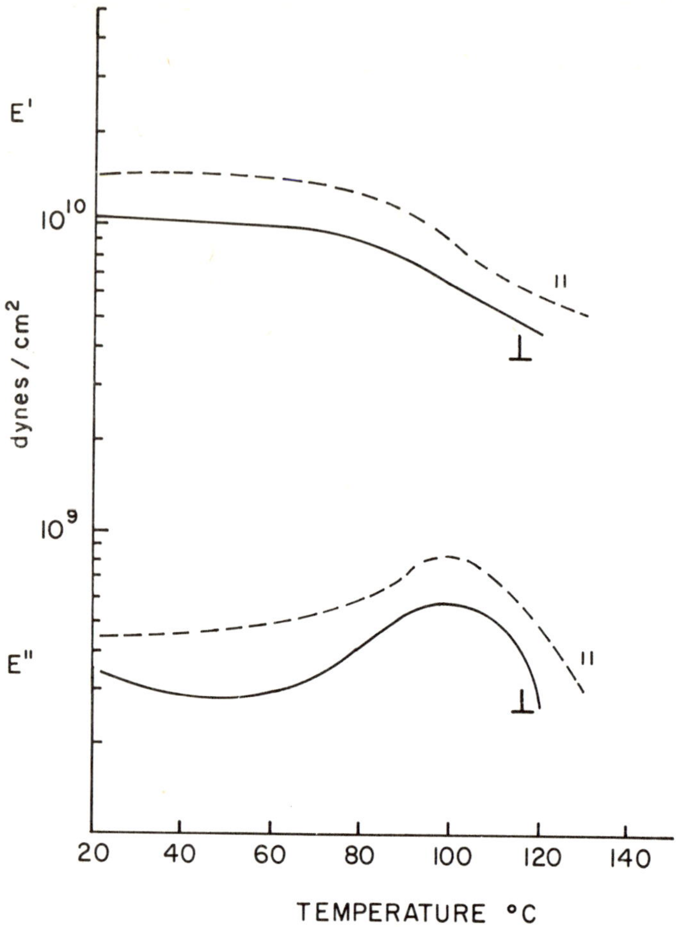

Fig. 6.1. Dynamic mechanical data taken parallel and perpendicular to the direction of stretch for nylon 66 film drawn 3X.

E", for samples taken along or perpendicular to the stretch direction, occurs at 90°C, and the E" curves are flatter than for the 3X sample. The E' curves cross at the temperature at which the maximum in E" occurs.

The results obtained for the 3X nylon 66 film are in agreement with the results on cold-drawn polyethylene and polypropylene [20]: the E' curves for samples parallel and perpendicular to the stretch direction do not cross (with E' parallel always greater than E' perpendicular). According to Takayanagi et al. [20], it was necessary to anneal the cold-drawn samples before the E' curves would cross. In his study, the annealing was presumably carried out during

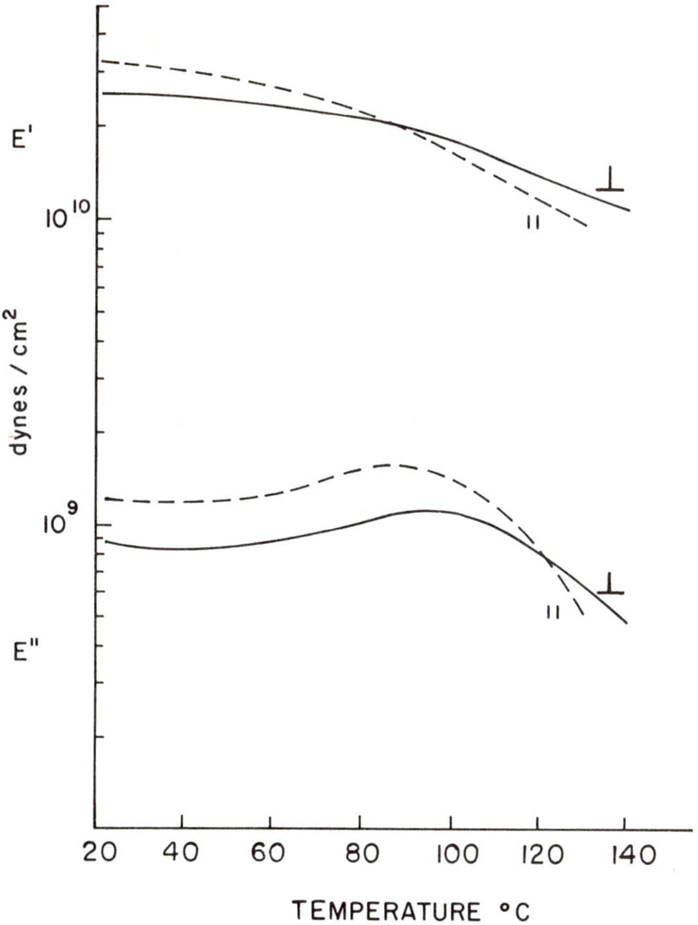

Fig. 6.2. Dynamic mechanical data taken parallel and perpendicular to the direction of stretch for nylon 66 film drawn 4.5X.

the drawing process for the 4.5X sample drawn at 200°C. This sample had E' values higher than the 3X sample, but the two curves (parallel and perpendicular) are close together, and since E' parallel to the stretch direction decreases faster with increasing temperature than E' perpendicular to the stretch direction, the curves cross. From these results, it seems that cold drawing causes the E' values to be well separated, whereas annealing causes them to be close together. The behavior of the E" curves is also of interest, for, although the E' values increase for the 4.5X samples, the position of the maximum of E" moves to lower temperatures, namely, 100°C and 90°C for the 3X and 4.5X samples. It is usually believed that E' is increased by an increase in orientation and that orientation causes the α-transition to occur at a higher temperature. Here, it is found that E' increases, while the α-transition moves to a lower temperature. This is a crystal-size effect, as is shown by small-angle x-ray measurements on annealed nylon 66 film, as reported in Table 6.1. Although

TABLE 6.1

Data for annealed nylon 66 films

Annealing temperature (°C)[a]	Density (g/cc)	Long period (Å)	Crystallite vol. v (10^3 Å^3)	Crystallinity χ(%)	10^7χ/
100	1.139	80.5	205	0.464	22.5
140	1.142	79.8	196	0.483	24.6
180	1.144	80.5	206	0.496	24.0
200	1.152	81.7	225	0.550	24.4
240	1.159	95.2	340	0.596	17.5

[a]6-hour anneal.

the orientation may increase throughout the series 3X and 4.5X, the number of crystals may actually decrease, thereby leading to a decrease in the temperature of the α-transition. A similar effect has been observed in PET [31]. Figure 6.3 shows the effect of orientation on nylon 66 yarn. An increase in the draw ratio causes the maximum of the E" curve to occur at a higher temperature. The E' curves are elevated as the draw ratio increases.

Orientation has more effect on the α-transition than does increase in crystallinity. As the draw ratio increases, the maximum in E" occurs at higher temperatures. Similar results were obtained by Thomas [32], who studied drawn nylon 66 up to a draw ratio of 4X using a torsion pendulum at a frequency of 0.3 Hz.

3. COMPRESSION

In recent years dynamic-compression mechanical analysis has been developed by modification of the Rheovibron [33,34] (see Chap. II). In this method compression grips are used on the two strain gauges. The compression grip is shown in Fig. 6.4. A small block or rod specimen is installed in the grip as shown. The compression grips consist of two sample mounting units and 13 cm long rods; the weight of the grips is 10.5 g. The tensile clamps are replaced by these compression grips at the connectors T_1 and T_7 of the strain gauge. One side of the sample specimen is fastened to the center of the T_7 sample mounting unit.

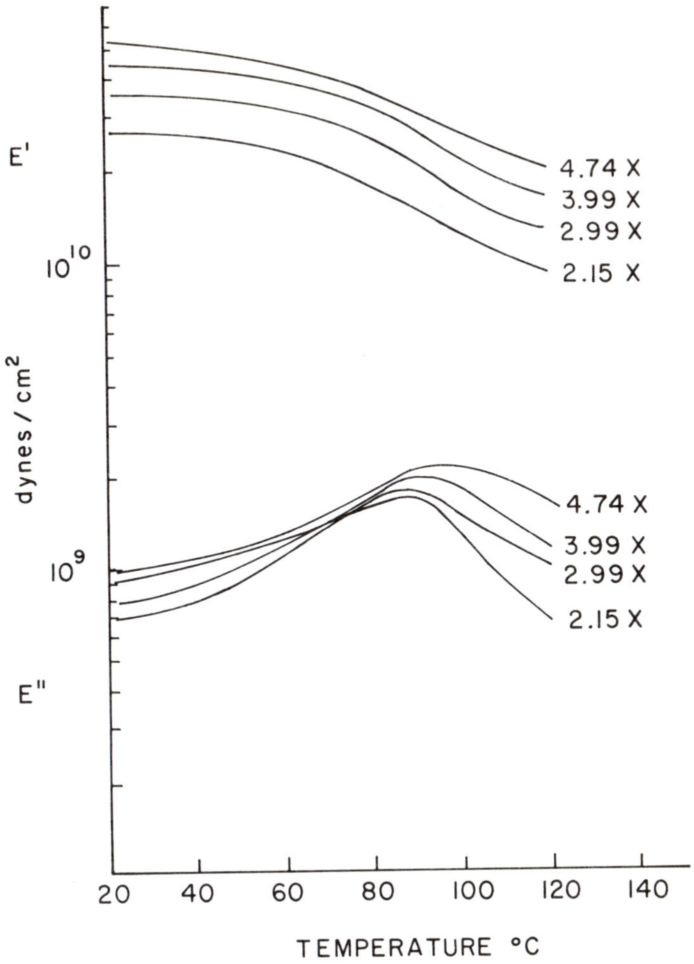

Fig. 6.3. Effect of drawing on the dynamic mechanical properties of nylon 66 fibers.

In order to calculate the dynamic compression modulus (E'_c), the following equation has been derived:

$$E'_c = \frac{2}{AD} \times 10^9 \times \frac{L}{S} \text{ dyn/cm}^2, \qquad (6.1)$$

where E'_c = dynamic compression modulus in dyn/cm², A = amplitude factor, D = reading of dynamic force dial, L = length of sample, and S = cross-section of sample.

By knowing the dynamic compression modulus (E'_c) and tan δ, the dynamic compression loss modulus (E''_c) can be determined as

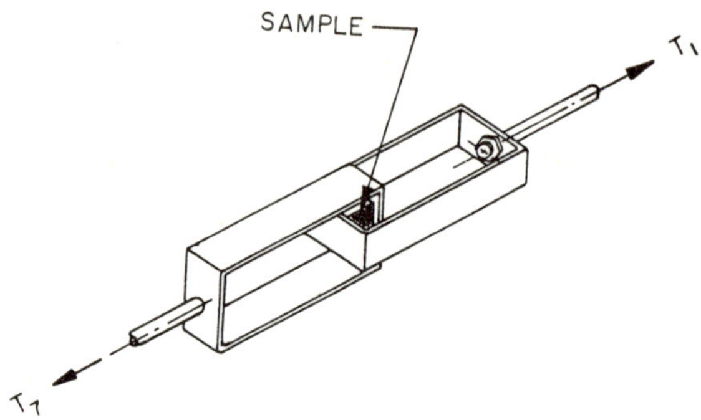

Fig. 6.4. Schematics of compression grip.

$$E''_c = (\tan \delta)\ (E'_c)\ \text{dyn/cm}^2. \tag{6.2}$$

A number of polyurethane elastomers, polybutadiene rubbers, and polyester sample blocks 0.6 X 0.4 cm and 0.4 cm thick were prepared. Dynamic measurements were made at 11 Hz with strain amplitude about 0.4 to 0.6%. Samples were heated at 1°C/min in nitrogen atmosphere, and measurements of the compression modulus E' and loss factor tan δ were made at 5 or 10°C increments. Samples were allowed to equilibrate at temperature for 10 minutes before measurements were made.

The dynamic compression modulus (E'_c), along with the loss tangent (tan δ) and the loss compression modulus, of a polyurethane elastomer are shown in Fig. 6.5. The dynamic compression modulus decreases with increasing temperature over a temperature range from -60°C to 100°C. The loss-tangent peak (T_g) of this elastomer is about -20°C. However, the intensity of the α-peak $(\tan \delta)_{max}$ is lower than in the tensile test.

Dynamic compression results have also been obtained on polybutadiene rubber. The T_g of this rubber is about -25°C. The compression modulus decreases sharply in the temperature range from -60 to -10°C.

Figure 6.6 gives the dynamic compression data on polyester film. The α-peak temperature of polyester film is about 96°C at 0% RH. This is in agreement with the results of the dynamic shear measurement on the same sample. However, the dynamic compression modulus for the sample is higher than dynamic shear modulus in the temperature

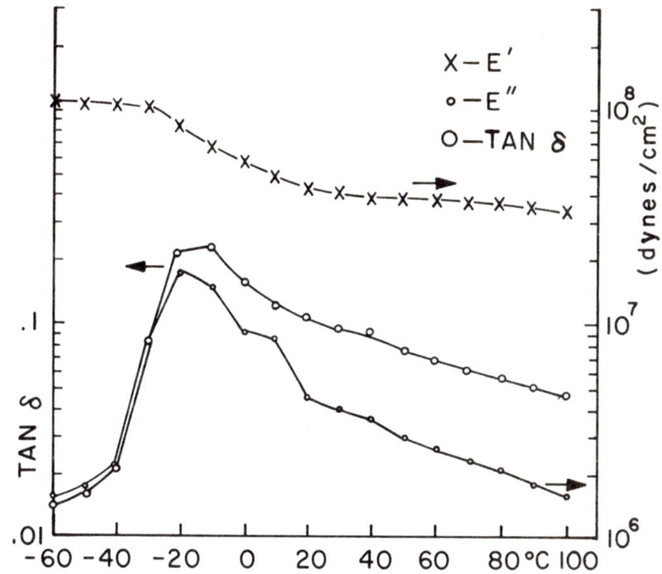

Fig. 6.5. Dynamic compression mechanical properties of polyurethane elastomer, 0% RH.

range of 23°C to 140°C. These results indicate that this polyester film has about 20% crystallinity and behaves anisotropically.

4. TORSION AND SHEAR

Torsion-pendulum and torsional-braid analysis are commonly used for measuring the dynamic shear modulus and mechanical damping of polymers. The specimen is a cylindrical rod or rectangular block. The oscillations twist and untwist the specimen, and dynamic response is measured. These techniques are discussed in Chap. II. A number of results from these techniques have been published [5, 35-37].

A modification of the Rheovibron for measurements on materials in the shear mode has been reported [38-40]. This technique uses new shear grips in the sample-clamping section as shown in Fig. 6.7, where two small specimen cubes are installed in the grips. Sufficient compressive force can be exerted by the adjustable side plates to contain most samples without slippage, since very small dynamic strain is involved. If slippage is troublesome, a small drop of polymer solvent may be applied to the surfaces of the sample to increase the friction. Once the sample is installed in the shear grips, the entire assembly can then be placed in the standard specimen

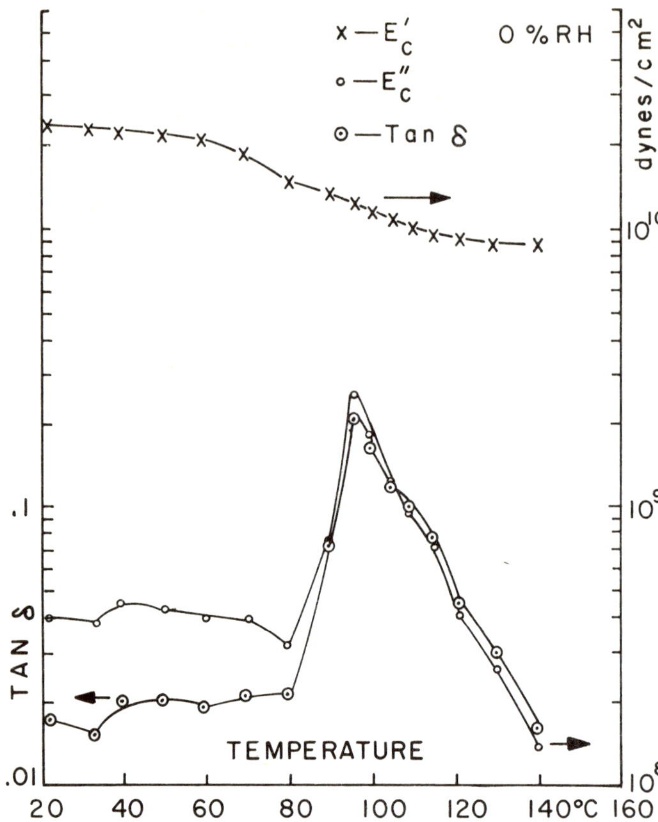

Fig. 6.6. Dynamic compression mechanical properties of polyester film.

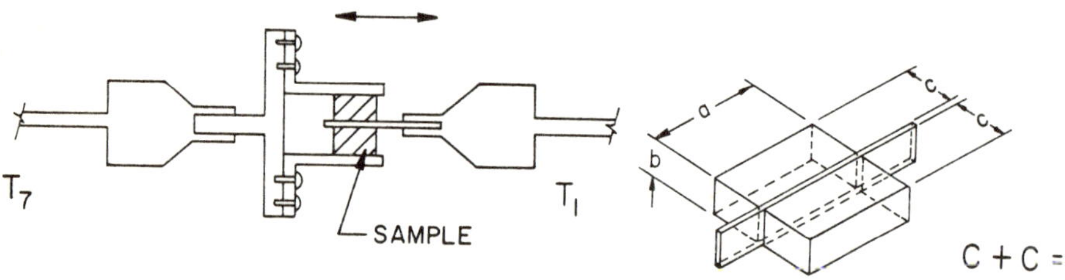

Fig. 6.7. Schematics of shear grips.

clips in the same manner as a tensile specimen. The back plate which holds the adjustable side plates should be held in the strain (T-7) clip, and the center shear plate should be held in the force (T-1) clip.

The following is the operating procedure for shear testing with the Rheovibron. With the tan δ range switch at 40 and amplitude factor switch at 30:

(1) Set main selector switch to stress (T-1) gauge position.
(2) Adjust knob until tan δ meter indicates zero.
(3) Set main selector switch to "Amp. F" position.
(4) Adjust "Amplitude Adjust" for full-scale indication on tan δ meter.
(5) Set main selector switch to "Dyn. F" position.
(6) Adjust "Phase Adjust" control for correct Lissajous pattern on oscilloscope (straight line at approximately 45° angle to horizontal).
(7) Adjust "Dynamic Force" potentiometer for full-scale indication on tan δ meter.
(8) Turn main selector switch to "tan δ" position and read tan δ.

In order to calculate the dynamic shear modulus (G'), the following equations have been derived [41]:

$$G' = \frac{\Delta F}{\Delta L} \frac{h}{4(ab)}, \tag{6.3}$$

$$G' = \frac{2}{AD} \times 10^9 \times \frac{h}{4(ab)} \text{ dyn/cm}^2, \tag{6.4}$$

where G' = dynamic shear modulus in dyn/cm^2, ΔF = oscillating load, ΔL = oscillating displacement, A = amplitude factor, D = reading of dynamic-force dial, h = width of two samples, a = length of sample, and b = thickness of sample.

By knowing the dynamic shear modulus (G') and tan δ value, the dynamic shear loss modulus (G") can be determined through the following relationship:

$$G'' = (\tan \delta)(G') \text{ dyn/cm}^2. \tag{6.5}$$

A number of polyurethane elastomer and polyester sample blocks, 0.6 X 0.5 cm and about 0.3 cm thick, were prepared. Dynamic measurements were made at 11 Hz with strain amplitude about 0.4 to 0.6%. Samples were heated at 1°C/min in nitrogen atmosphere under relaxed conditions, and measurements of the shear modulus G' and loss factor tan δ were made at 5 or 10°C increments. Samples were allowed to equilibrate at temperature for 10 minutes before measurements were made.

The dynamic shear modulus (G') along with the loss tangent (tan δ) and the loss shear modulus (G") of polyurethane elastomer are shown in Fig. 6.8. The α-peak temperature of this polymer is about -34°C. The dynamic shear modulus decreases with increasing temperature.

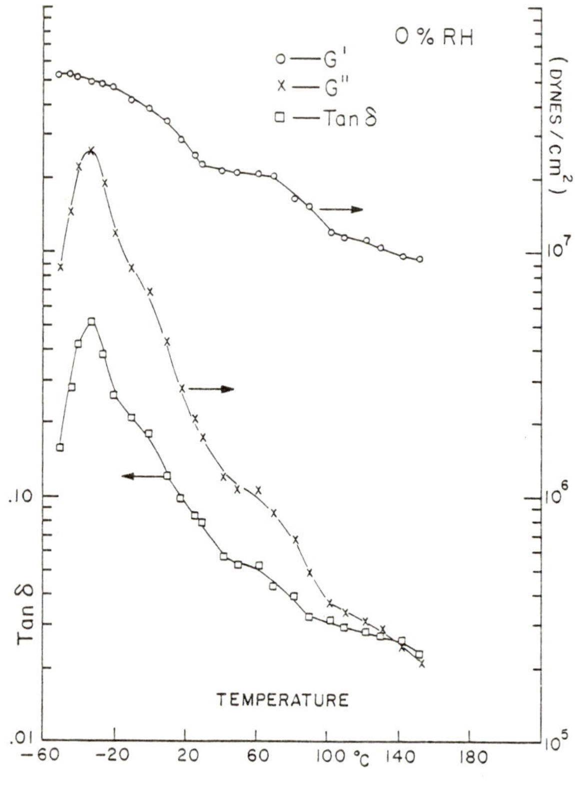

Fig. 6.8. Dynamic shear mechanical properties of polyurethane elastomer.

Figure 6.9 gives the dynamic shear data on polyester film. The α-peak temperature is about 92°C at 0% RH. This result suggests that this polyester film has about 20% crystallinity. Thus the new shear grip and procedure make it possible to measure dynamic mechanical properties in the shear mode using the Rheovibron Viscoelastometer.

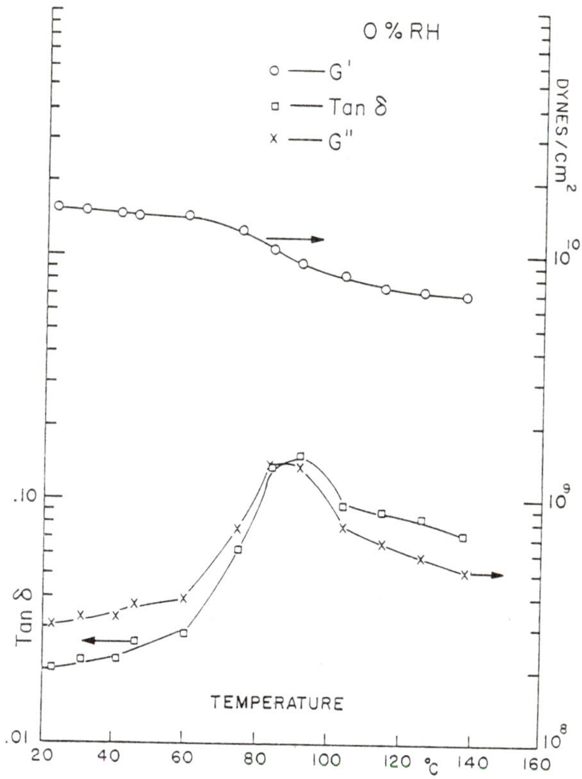

Fig. 6.9. Dynamic shear mechanical properties of polyester film.

5. BENDING

Dynamic bending mechanical analysis is useful for obtaining the transverse properties of materials. The B&K complex-modulus apparatus and several mechanical spectrometers are equipped to obtain the dynamic properties in bending mode. These instruments are described in Chap. II.

In this section a new method for investigating dynamic bending mechanical properties of materials is presented, as well as results on the effects of temperature on dynamic bending properties of polyester, nylon, and Acrilan fiber [42].

This method uses a modification of the Rheovibron Viscoelastometer [43] by the installation of bending grips as shown in Fig. 6.10. These consist of one sample-mounting unit and clamp with 15-cm-long rods; the weight of the grips is 13.7 g. The tensile clamp is

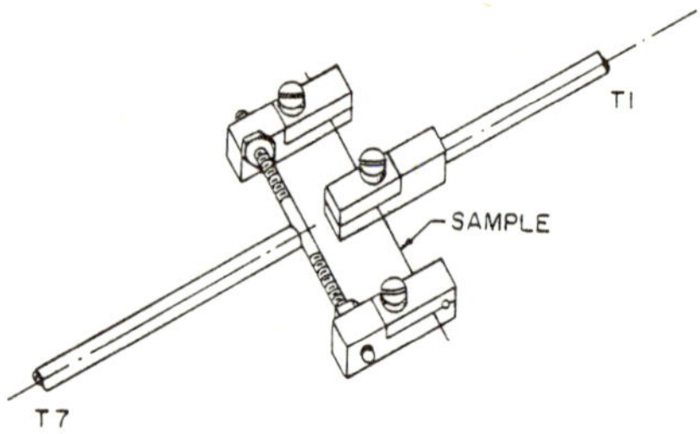

Fig. 6.10. Schematic of bending grip.

replaced by these bending grips at the connectors T_1 and T_7 of the strain gauge. The sample is mounted vertically in the T_7 clamp. In case of film or monofil sample, the T_1 clamp is used to fasten the center of the sample. In measurement of bundle yarn, the end of the T_1 clamp is replaced by a latched needle in order to avoid flattening the yarn.

The operating procedure for bending testing with the Rheovibron is as follows. With the tan δ range switch at 30 and the amplitude factor at 30:

(1) Set main selector switch to stress (T-1) gauge position.
(2) Turn handle of slider of driving section and apply initial bending, about 0.2-0.5 g to sample.
(3) Set main selector switch to "Amp. F" position, and turn tan δ meter switch to 40 and amplitude factor switch to 30 or 20.
(4) Adjust "Amplitude Adjust" for full-scale indication on tan δ meter.
(5) Set main selector switch to "Dyn. F" position.
(6) Adjust "Phase Adjust" control for correct Lissajous pattern on oscilloscope.
(7) Adjust "Dynamic Force" potentiometer for full-scale indication on tan δ meter.
(8) Turn main selector switch to "tan δ" position and read tan δ.

In order to calculate the dynamic bending modulus (E'_B), the following equation has been derived:

$$E'_B = \frac{2}{AD} \times 10^9 \times \frac{L^3}{96\pi^4}, \qquad (6.6)$$

where E'_B = dynamic bending modulus in dyn/cm², A = amplitude factor, D = reading of dynamic-force dial, L = length of sample, and γ = radius of sample.

By knowing the dynamic bending modulus E'_B and tan δ, the dynamic bending loss modulus E''_B can be determined from the following relationship:

$$E''_B = (\tan\delta)(E'_B) \text{ dyn/cm}^2. \quad (6.7)$$

A number of polyester, nylon 66 monofil, and acrylic fibers were prepared. Dynamic measurements were made at 11 Hz with strain amplitude about 0.5%. Samples were heated at 1°C/min in nitrogen atmosphere, and measurements of the bending modulus E'_B and damping factor tan δ were made at 5 or 10°C increments. Samples were allowed to equilibrate at temperature for 15 minutes before measurements were made.

The dynamic bending modulus (E'_B), along with the loss tangent (tan δ) and the loss bending modulus, of nylon 66 monofilament are shown in Fig. 6.11. The dynamic bending modulus decreases with

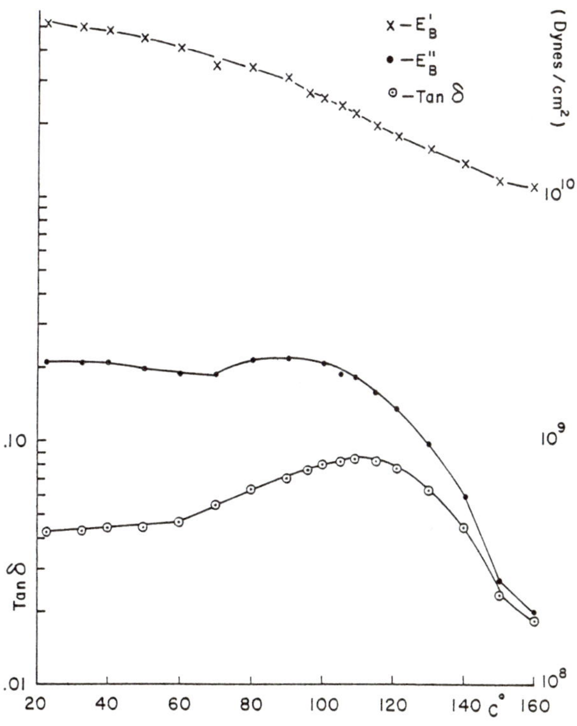

Fig. 6.11. Dynamic bending mechanical properties of nylon 66 monofilament (2060 denier, 4.5X).

increasing temperature in the range from 50°C to 160°C. The loss-tangent peak (T_g) of this nylon 66 is about 110°C. However, the intensity of the α-peak, $(\tan \delta)_{max}$, is slightly lower than in the tensile test.

The dynamic-bending results for polyester monofilament and yarn are shown in Figs. 6.12 and 6.13. The T_g of these polyester samples

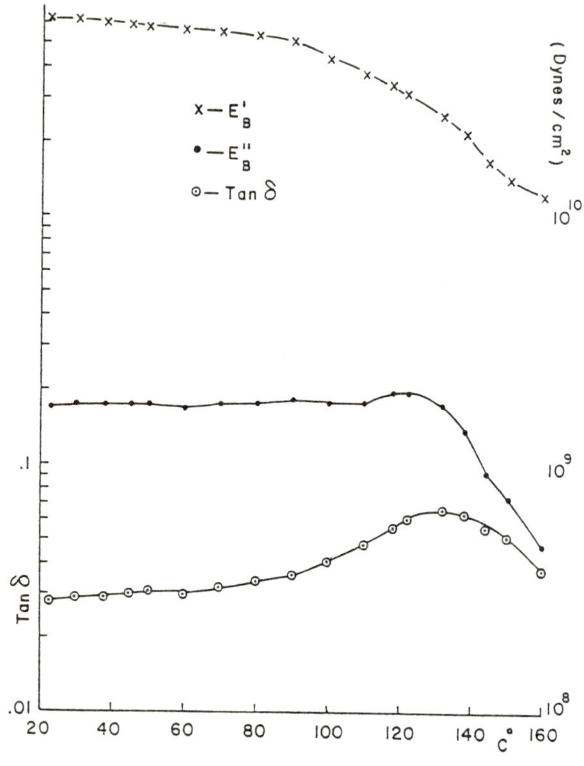

Fig. 6.12. Dynamic bending mechanical properties of PET monofilament (2010 denier, 4.5X).

is about 130°C. The dynamic bending modulus is lower than the tensile modulus in the temperature range from 25 to 160°C. The α-peak temperature is in agreement with the dynamic tensile measurement. However, the value of the loss tangent in bending is lower than in the tensile measurement. Figure 6.14 shows the results on acrylic fiber. The dynamic bending modulus decreases sharply in the temperature range from 60°C to 160°C.

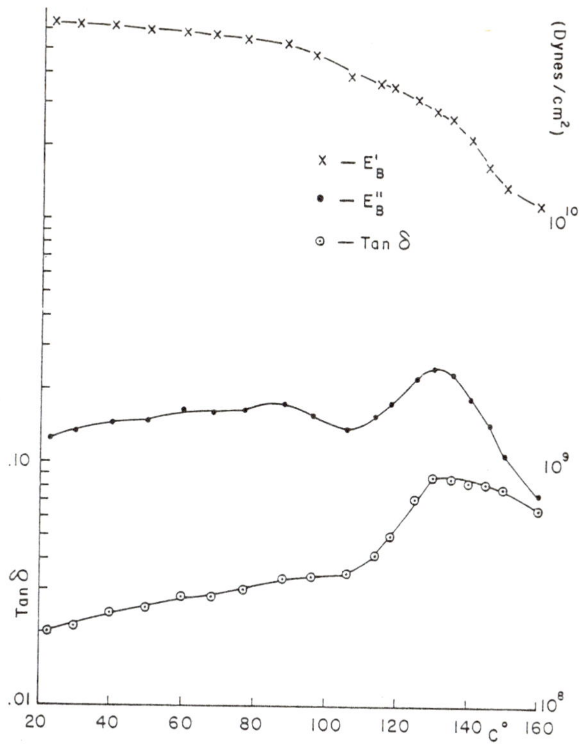

Fig. 6.13. Dynamic bending mechanical properties of polyester yarn (1150/230, 5.0X).

This new method of measurement of dynamic bending properties is useful for analyses in connection with end-use performance of fiber structure.

REFERENCES

1 J. F. Nye, Physical Properties of Crystals, Their Representation by Tensors and Matrices, Oxford U. P., 1957.
2 I. M. Ward, Mechanical Properties of Solid Polymers, Wiley-Interscience, New York, 1971.
3 R. F. Hearmon, Introduction to Applied Anisotropic Elasticity, Oxford U. P., 1961.
4 J. E. Ashton and J. M. Whitney, Theory of Laminated Plates, Technomic, Stamford, CT, 1970.
5 L. E.Nielsen, Mechanical Properties of Polymers and Composites, Marcel Dekker, New York, 1974.
6 G. Raumann, Proc. Phys. Soc., 79 (1962) 1221.
7 G. Raumann and D. W. Saunders, Proc. Phys. Soc., 77 (1961) 1028.
8 P. R. Pinnock and I. M. Ward, Proc. Phys. Soc., 81 (1963) 260.
9 I. M. Ward, Textile Res. J., 31 (1961) 650.

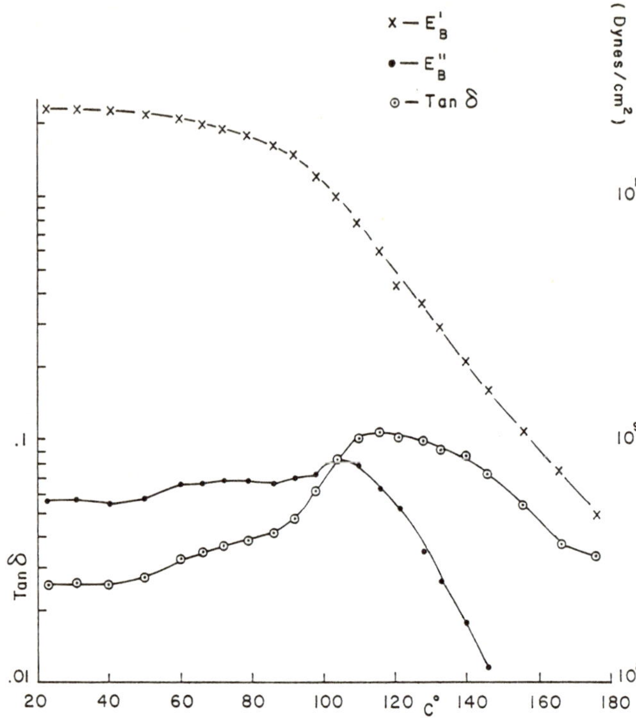

Fig. 6.14. Dynamic bending mechanical properties of acrylic fibers.

10 D. W. Hadley, I. M. Ward, and J. Ward, Proc. R. Soc., A285 (1965) 275.
11 P. R. Pinnock, I. M. Ward, and J. M. Wolfe, Proc. R. Soc., A291 (1966) 267.
12 D. W. Hadley, P. R. Pinnock, and I. M. Ward, J. Mater. Sci., 4 (1969) 152.
13 R. Meredith, J. Textile Inst., 45 (1954) 489.
14 H. Kolsky and K. W. Hillier, Proc. Phys. Soc., B62 (1949) 111.
15 J. W. Ballou and J. C. Smith, J. Appl. Phys., 20 (1949) 493.
16 H. M. Morgan, Textile Res. J., 32 (1962) 866.
17 W. J. Hamburger, Textile Res. J., 18 (1948) 705.
18 W. H. Church and W. W. Moseley, Textile Res. J., 29 (1959) 525.
19 W. W. Moseley, J. Appl. Polym. Sci., 3 (1960) 266.
20 M. Takayanagi, K. Imada, and T. Kajiyama, J. Polym. Sci., C-15 (1966) 263.
21 Z. H. Stachurski and I. M. Ward, J. Polym. Sci., A-2, 6 (1968) 1083.
22 V. B. Gupta and I. M. Ward, J. Macromol. Sci. (Phys.), B2 (1968) 89.
23 I. L. Hay and A. Keller, J. Mater. Sci., 1 (1966) 41.
24 T. Seto and T. Hara, Rep. Progr. Polym. Phys. (Jap.), 7 (1963) 63.
25 Z. H. Stachurski and I. M.Ward, J. Polym. Sci., A-2, 6 (1968) 1817.

26 Z. H. Stachurski and I. M. Ward, J. Macromol. Sci. (Phys.), B3 (3) (1969) 445.
27 J. H. Dumbleton and T. Murayama, Kolloid-Z. Z.-Polym., 238 (1970) 411.
28 A. E. Woodward, J. A. Sauer, C. W. Deeley, and D. E. Kline, J. Coll. Sci., 12 (1957) 363.
29 A. E. Woodward, R. E. Glick, J. A. Sauer, and R. P. Gupta, J. Polym. Sci., 35 (1960) 367.
30 A. E. Woodward, J. M. Crissman, and J. A. Sauer, J. Polym. Sci., 34 (1960) 23.
31 J. H. Dumbleton, J. P. Bell, and T. Murayama, Kolloid-Z. Z.-Polym., 228 (1968) 54.
32 A. M. Thomas, Nature, 179 (1957) 862.
33 T. Murayama, J. Appl. Polym. Sci., 19 (1976) 2593.
34 T. Murayama, Proceedings on the 7th International Congress on Rheology, Gothenburg, Sweden, 1976, 402.
35 L. E. Nielsen, Rev. Sci. Instrum., 22 (1951) 690.
36 J. K. Gillham, Polym. Eng. Sci., 7 (1967) 225.
37 B. Hartman and G. F. Lee, J. Appl. Polym. Sci., 21 (1977) 1341.
38 D. J. Massa, J. Appl. Phys., 44 (1973) 2595.
39 P. F. Erhardt, J. J. O'Malley, and R. G. Crystal, in Block Copolymers (S. L. Aggarwal, Ed.), Plenum, New York, 1970, p. 195.
40 T. Murayama, J. Appl. Polym. Sci., 19 (1975) 3221.
41 S. Timoshenko, Strength of Materials, Part 1, Elementary Theory and Problems, Van Nostrand, Princeton, NJ, 1955.
42 T. Murayama, Proceedings of 26th International Congress of Pure and Applied Chemistry, Tokyo, 1977, p. 143.
43 M. Takayanagi, Mem. Fac. Eng., Kyushu Univ., 23 (1963) 1.

CHAPTER VII

DYNAMIC PROPERTIES IN GAS AND LIQUID MEDIA

1. INTRODUCTION

External variables which affect dynamic mechanical properties of polymers include various deformations (shear, tensile, bending, etc.), the temperature, and the geometry of structural assemblies. Effects of these variables on dynamic properties are discussed in previous chapters. Properties of the environment (liquid or gas) are also important in determining the dynamic mechanical properties of polymers, which affect both end-use performance and process control.

In this chapter, the techniques of dynamic mechanical analysis in gas and liquid media will be discussed, along with related subjects: the effects of humidity, relationships between dynamic loss modulus and diffusion, and dynamic mechanical studies on biopolymers.

2. IN GAS

In recent years, a new method for investigating dynamic mechanical properties of materials in a gaseous medium has been reported [1]. This apparatus and procedure has been useful in studies on the effect of different gases on polymers in use, and on the effects of gas treatment on materials. The apparatus was developed by Murayama and Silverman using the Rheovibron instrument. The new gas cell used will also be adaptable to other mechanical testing instruments such as the Instron, Tensometer, and torsion pendulum.

To measure the dynamic properties in a gas medium it was necessary to devise an apparatus to keep the sample immersed in gas and the transducers outside the gas, yet aligned with the sample. This was accomplished by using a sample gas cell with two liquid seals. The driving unit, T_7 strain gauge, sample gas cell, T_1 strain gauge, and slider were mounted vertically on a stand. A sketch of the apparatus is shown in Fig. 7.1. The amplifier with oscillator and the stabilized eliminator are connected to strain gauges and the driving unit.

The sample gas cell made of glass, is 6 inches in diameter and $9\frac{3}{4}$ inches long. The upper part of the cell has a $\frac{1}{2}$-inch-diameter, 1-inch-high glass tube to admit the sample clamp; an oil-deposit

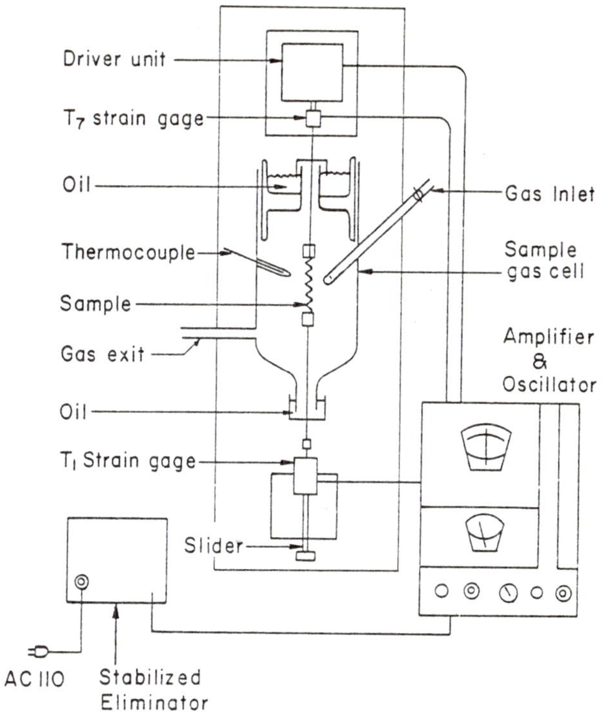

Fig. 7.1. Schematic of apparatus.

portion; and a tight ground-glass seal about $1\frac{1}{8}$ inch in diameter. The lower part of the cell consists of a $\frac{1}{4}$-inch-diameter glass-tube gas inlet; a $\frac{1}{4}$-inch-diameter, 1-inch-long gas outlet; a $\frac{1}{2}$-inch-diameter, 1-inch-long sample-clamp inlet; and a $\frac{1}{4}$-inch-diameter, 1-inch long thermocouple inlet. The gas cell is shown in Fig. 7.2 (A). The volume of the gas cell is about 1 liter.

In order to insert a sample into the cell without loss of gas from the inlet of the sample clamp, the liquid seal with clamp rod is designed as shown in Fig. 7.2(B). The clamp rod is connected with the liquid seal, which is made from a polyethylene cap ($1\frac{3}{8}$-inch diameter, 1-inch height). One pair of clamp rods with liquid seal is passed through a $\frac{1}{2}$-inch-diameter inlet in the center of the gas cell, as shown in Fig. 7.2. The weight of the clamp with rod is controlled at 11 g. The total weight of the clamp and seal cap with oil is about 25 g. Fisher paraffin oil (15 cc) is used as an oil seal. The circuit of the Vibron is modified as shown in Fig. 7.3. A new controller (E, F) is built to adjust the resistor of T_1 and

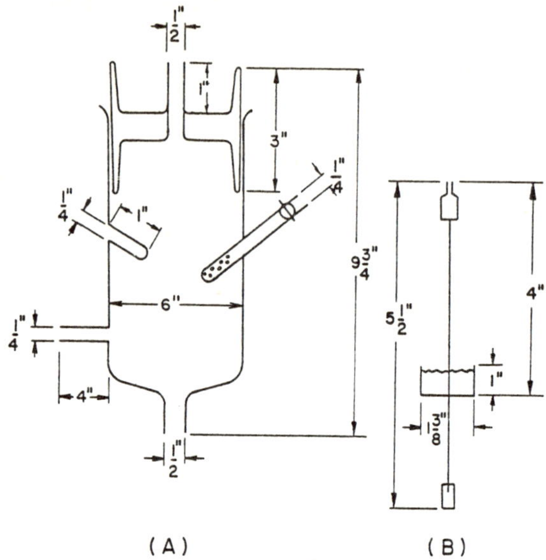

Fig. 7.2. Gas cell (A) and clamp with liquid seal (B).

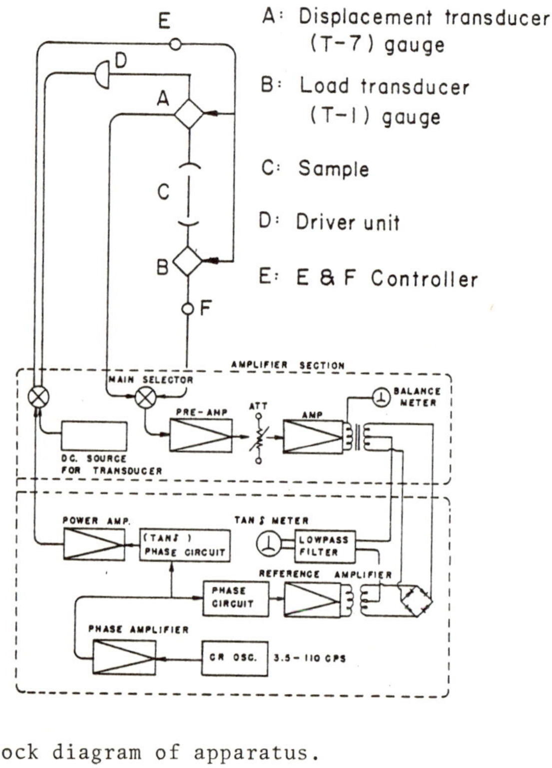

Fig. 7.3. Block diagram of apparatus.

T_7 gauge, since the arrangement of the driving unit and clamps is changed.

To demonstrate the apparatus and procedure, the dynamic mechanical properties of dry nylon 66 filaments were measured in three different gas media (100, 50-50, and 25-75 HCl-N_2 gas ratios) at 11 Hz as a function of time at room temperature (25°C).

For the immersion of the test sample and the gas control system, the apparatus is placed in a hood to avoid the health hazard of HCl gas. The gas concentration is adjusted by the gas control system, as shown in Fig. 7.4. Data on the dynamic modulus are shown in

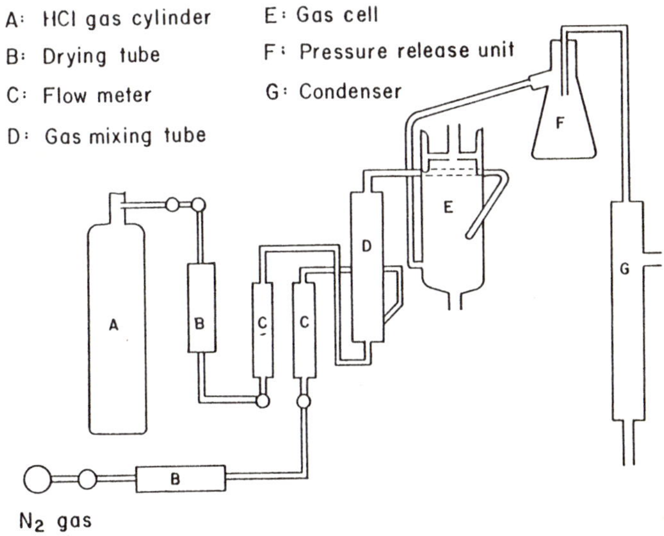

Fig. 7.4. Gas control system for sample gas cell.

Fig. 7.5, and on the loss tangent (tan δ) in Fig. 7.6. These results indicate that the tan δ and dynamic modulus (E') of dry nylon fiber gassed to equilibrium were the same at HCl concentrations of 25, 50, and 100%. Short-time gassing data indicate that the HCl sorption by nylon 66 is exothermic, and equilibrium sorption of HCl by nylon 66 doubles tan δ and reduces the dynamic modulus (E') by 20%.

3. IN LIQUID

(1) Apparatus; effect of liquid viscosity and frequency. Numerous studies of dynamic mechanical properties have been made in the temperature range from -160°C to 250°C in an air atmosphere at 0% relative humidity. A supplemental chamber has been designed to provide an environment of either 0 or 100% relative humidity. Studies at 100%

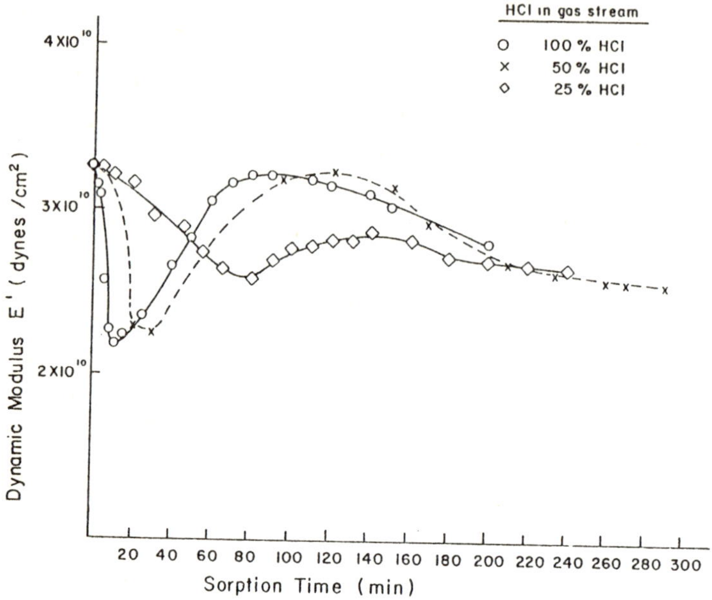

Fig. 7.5. Effect of HCl sorption on the dynamic modulus of nylon 66 fibers.

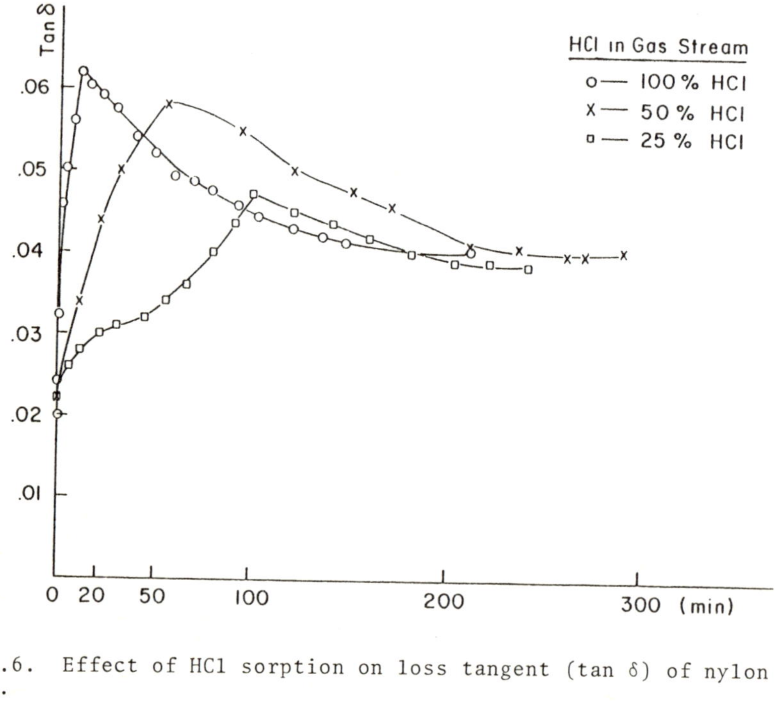

Fig. 7.6. Effect of HCl sorption on loss tangent (tan δ) of nylon 66 fibers.

RH have been reported by several investigators [2-4]. Measurements of dynamic mechanical properties of materials in a liquid medium using the Rheovibron instrument are discussed in this section [5].

The Rheovibron applies a sinusoidal tensile strain to one end of the sample and measures the stress output at the other end. The instrument operates at frequencies of 3.5, 11, 35, and 110 Hz. Two transducers are used to read directly the absolute dynamic modulus $|E^*|$ (the ratio of maximum stress amplitude to maximum strain amplitude) and the phase angle δ between stress and strain. From these two quantities, the real part E' (dynamic modulus) and the imaginary part E" (loss modulus) can be calculated from the dynamic tensile modulus |E"|. The principles of this direct-reading method and instrument are described in detail by Takayanagi [6].

To measure the dynamic properties in a liquid medium, it was necessary to devise an apparatus to keep the sample immersed in liquid and the transducers outside the liquid, yet aligned with the sample. This was accomplished by using two liquid baths. Two horizontally opposed $\frac{1}{4}$-inch-diameter holes were drilled in the walls of the sample bath to accommodate the two rods between the sample clamps and transducers. The liquid medium was pumped from another bath into the sample bath at a flow rate which maintained the liquid level 1 inch above the sample, clamps, and rods. A sketch of the apparatus is shown in Fig. 7.7.

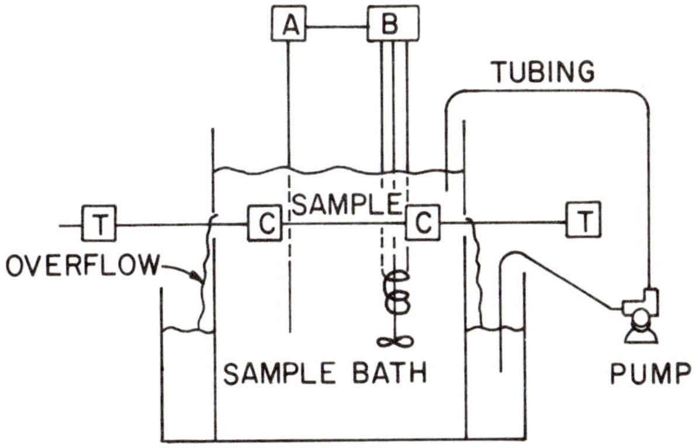

Fig. 7.7. Experimental arrangement for a liquid medium. A, thermoregulator; B, circu-temp temperature controller; C, clamps; T, transducers.

The sample bath was made from a $5\frac{3}{4}$-inch-diameter glass beaker, and the second bath from a $7\frac{1}{2}$-inch-diameter beaker. The height of the sample bath was 6 inches, and of the second bath 3 inches. Two horizontally opposed $\frac{1}{4}$-inch-diameter holes were drilled in the sample bath 2 inches from the top. The two rods and clamps were placed through these holes and the fiber sample attached to the two clamps. The liquid medium was placed in both baths. A Fisher suction pump was used to circulate the liquid medium at a flow rate (500 ml/min) sufficient to maintain a liquid level 1 inch above the sample. A thermoregulator and circu-temp temperature controller (which contained a stirrer and electrical heating element) were used to maintain a constant temperature to ±1°C in the sample bath.

The importance of this type of experiment can be readily seen in that some polymeric solids change properties in the presence of a liquid. This can influence both the processing and the end use of the material. One needs to be able to distinguish what parts of the dynamic modulus and loss tangent (tan δ) are due to the intrinsic properties of the solid, and what parts are due to the drag forces of the liquid on the surface of the solid. A theoretical and experimental investigation of the dynamic mechanical properties of polymeric films in liquid media [7] shows the effects of viscous drag on the dynamic mechanical properties in the presence of a liquid. In this study a mathematical analysis is carried out for a film submerged in liquid of density ρ, viscosity η, and kinematic viscosity $\nu = \eta/\rho$, with a forced sinusoidal oscillation at one end ($x = L$) and fixed at the other ($x = 0$). The assumptions employed in the analysis are as follows: (1) there is no flow into or out of the sheet and no chemical reaction with the liquid; (2) the free surface of the liquid is unaffected by the oscillating film; (3) the liquids are Newtonian; (4) the steady-state solution will be used; (5) variations in the z-direction (the direction of the width of the sample) are neglected; (6) variations in cross-sectional and surface areas are neglected. Under these assumptions the Navier-Stokes equations simplify to

$$\frac{\partial u}{\partial t} = \nu \frac{\partial^2 u}{\partial y^2}, \tag{7.1}$$

where u = velocity of the liquid in the x direction, t = time, and y = coordinate perpendicular to the film. The boundary conditions are

$$\text{at } y = o,\ u = \text{Re}\ U_0 e^{i(kx-\omega t)} = \frac{\partial s}{\partial t},$$

where s = displacement in the film, U_0 = complex velocity, u = amplitude, ω = frequency; and

$$\text{at } y = \infty,\ u = 0.$$

The steady-state solution is

$$u = U(y)e^{i(Kx-\omega t)}, \tag{7.2}$$

where

$$U(y) = U_0 e^{-\lambda y} \tag{7.3}$$

and

$$\lambda = (-i\omega/\nu)^{1/2}. \tag{7.4}$$

Now a force balance on the film leads to

$$\rho_f \frac{\partial^2 s}{\partial t^2} = E \frac{\partial^2 s}{\partial x^2} - \frac{2\mu\lambda}{\rho_f h}\frac{\partial s}{\partial t}, \tag{7.5}$$

where ρ_f is the film density, h is the film thickness, and E is the complex dynamic modulus of the dry film, a function of frequency only. By substituting

$$s = s_0 e^{i(kx\ -\ \omega t)}, \tag{7.6}$$

one obtains

$$k^2 = \frac{1}{E}\left[\rho_f\omega^2 + \frac{2i\mu\omega}{h}\left(\frac{\omega}{2\nu}\right)^{1/2} + \frac{2\mu\omega}{h}\left(\frac{\omega}{2\nu}\right)^{1/2}\right]. \tag{7.7}$$

Now

$$s = \text{Re } a\, \frac{\sin kx}{\sin kL}\, e^{-i\omega t}, \tag{7.8}$$

where a is the displacement amplitude, and since $|kx|$ is small,

$$\begin{aligned}\sigma_0 &= E\left(\frac{\partial s}{\partial x}\right)\ x = o\\ &= \frac{a}{L}\,\text{Re } E\left[1 + \frac{k^2L^2}{6}\right] e^{-i\omega t}.\end{aligned} \tag{7.9}$$

Then the measured stress is

$$\begin{aligned}\sigma_0 = &\left[E' + \frac{L^2\rho_f\omega^2}{6} + L^2\frac{\mu\omega}{3h}\left(\frac{\omega}{2\nu}\right)^{1/2}\right]\cos\omega t\\ &+ \left[E'' + \frac{L^2\mu\omega}{3h}\left(\frac{\omega}{2\nu}\right)^{1/2}\right]\sin\omega t,\end{aligned} \tag{7.10}$$

where L is the length of the sample, and ν is the kinematic viscosity of the liquid. The stress can also be written as

$$\sigma_0 = (E' + k')\cos\omega t + (E'' + k'')\sin\omega t, \qquad (7.11)$$

where

$$k' = \frac{L^2\rho_f\omega^2}{6} + \frac{L^2\mu\omega}{3h}\left(\frac{\omega}{2\nu}\right)^{1/2}, \qquad (7.12)$$

$$k'' = \frac{L^2\mu}{3h}\,\omega\left(\frac{\omega}{2\nu}\right)^{1/2}. \qquad (7.13)$$

The effective dynamic modulus, E'_c, and the effective loss modulus, E''_c, are

$$E'_c = E'_{dry} + k', \qquad (7.14)$$

$$E''_c = E''_{dry} + k''. \qquad (7.15)$$

According to the theory advanced in the mathematical analysis, the effective dynamic modulus E'_c and loss modulus E''_c in liquids can be calculated if one knows the dynamic and loss modulus in the dry sample, the thickness and density of the sample, the viscosity of the liquid, and the frequency [equations (7.14) and (7.15)]. The predicted effective dynamic and loss moduli and loss tangent [$\tan\delta_c = (E''_{dry} + k'')/(E'_{dry} + k')$] of both films in the different liquid media are shown in Tables 7.2 and 7.3 below. The parameters k' and k" for predicting effective dynamic and loss modulus in various liquids are shown in Table 7.3. Reasonable agreement is observed between these predicted dynamic mechanical parameters (E'_c, E''_c, and $\tan\delta_c$) and those determined from experimental dynamic mechanical measurements in the different viscous liquids. However, while the magnitude and direction of the change are correct, the predicted values are low. This is thought to be a result of neglecting changes in the liquid in the z-direction (calculations for the cylindrical case support this view) and also to neglecting the presence of the container walls.

In view of this approximate agreement between the measured results and those predicted by the theory, the following investigation was made into the effect of the viscous drag on the dynamic mechanical properties in the presence of a liquid.

In order to obtain the effects of viscous drag on the dynamic mechanical properties, polymeric films and liquids which are known not to interact were prepared. The films are polyethylene (A, B) and polytetrafluoroethylene (Teflon) (C, D) [7]. These films were investigated in five different liquids (water, ethylene glycol, glycerine, and silicone fluids 1 and 2).

The viscosity and density of these liquids and the dimensions of test samples are shown in Table 7.1. Measurements of dynamic

TABLE 7.1
(a) Viscosity and density of various liquids

Liquid	Viscosity (cp)	Density (g/cc)
H_2O	0.89	1.0
Ethylene glycol	16.2	1.15
Glycerine	950	1.60
Silicone fluid 1	12,000	0.96
Silicone fluid 2	57,600	0.96

(b) Sample dimensions

Polyethylene	Density	Length (cm)	Width (cm)	Thickness (cm)	Test liquid
(A)	0.94	5	0.5	0.003	H_2O, ethylene glycol, glycerine
(B)	0.92	5	0.3	0.01	Silicone fluid 1 and 2

properties, dry and in liquids, were made at 25°C. A total of 10 tests on 3 samples of each material in each liquid were made at constant frequencies of 11 and 110 Hz with a Rheovibron instrument.

Tables 7.2 and 7.3 give the average values of the measured dynamic modulus (E'_e), loss modulus (E''_e), and loss tangent ($\tan \delta_e$) for the films in the five different liquid media. The results indicate that both moduli and loss tangent increased slightly with increasing liquid viscosity and frequency of oscillation. However, the differences between the dynamic modulus measured dry and in these liquids are very small.

The following conclusions may be drawn:

(1) The expressions deduced are demonstrated to predict the effective dynamic mechanical properties of films in liquid media from a knowledge of dimensions and density of the sample, the viscosity of the liquid, and the frequency. The effective dynamic and loss moduli increase slightly with increasing liquid viscosity and frequency of oscillation.

(2) The effect of viscous drag on the dynamic properties is small except in a very viscous liquid at high frequencies for a film of low modulus. Thus, within the capability of the Rheovibron, larger differences between the dynamic properties measured dry and

TABLE 7.2

Predicted and measured dynamic modulus, loss modulus, and loss tange of polyethylene films in various liquids

Conditions	Freq. (Hz.)	Dynamic modulus (10^9 dyn/cm^2) E'_e	E'_c	Loss modulus (10^8 dyn/cm^2) E''_e	E''_c	Loss tangent tan δ_e	tan δ
Dry PE (A)	11	6.433	--	5.983	--	.093	--
	110	6.475	--	5.957	--	.092	--
H_2O	11	6.434	6.433	5.984	5.984	.093	.093
	110	6.488	6.480	6.228	5.991	.096	.092
Ethylene glycol	11	6.435	6.434	6.306	5.983	.098	.093
	110	6.537	6.493	6.863	6.115	.105	.094
Glycerine	11	6.450	6.437	7.224	6.027	.112	.094
	110	6.885	6.616	7.435	7.347	.108	.111
Dry PE (B)	11	5.574	--	5.351	--	.096	--
	110	5.586	--	5.195	--	.093	--
Silicone fluid 1	11	5.593	5.578	6.432	5.387	.115	.097
	110	6.035	5.703	7.242	6.345	.120	.111
Silicone fluid 2	11	5.604	5.655	6.725	5.431	.120	.096
	110	6.448	5.838	8.060	7.695	.125	.132

TABLE 7.3

Factors k' and k" for predicting effective dynamic and loss modulus of polyethylene in various liquids

Conditions	Freq. (Hz.)	k' (dyn/cm^2)	k" (dyn/cm^2)
H_2O	11	1.25×10^5	1.06 X 10
	110	5.23×10^6	3.35 X 10
Ethylene glycol	11	5.07×10^5	4.88 X 10
	110	1.77×10^7	1.58 X 10
Glycerine	11	4.42×10^6	4.40 X 10
	110	1.41×10^8	1.39 X 10
Silicone fluid 1	11	3.65×10^6	3.63 X 10
	110	1.17×10^8	1.15 X 10
Silicone fluid 2	11	8.00×10^6	7.98 X 10
	110	2.52×10^8	2.50 X 10

in liquid will be seen if the liquid interacts chemically or physically with the film.

As a result of these studies the capability of dynamic mechanica analysis has been expanded to include investigation of the dynamic properties of polymers in liquids as a function of temperature [8-11]. The dynamic mechanical properties of nylon 66 fiber in two different liquid media (deionized water and tetrachloroethylene) have been reported using this new apparatus and procedure [5]. The apparatus is placed in a hood for immersion of the test sample, and a dome with a condenser is used above the bath to avoid the health hazard of chlorinated solvents. Samples of nylon (840 denier, 140-filament bundle) were heated at a rate of 1°C/min in a liquid medium in a relaxed condition, and measurements of the tensile modulus and loss tangent (tan δ) were made at 5 or 10°C increments. Samples were allowed to equilibrate at each temperature for 20 minutes before measurements were made.

Figures 7.8 and 7.9 show the result of a determination of the dynamic properties (E', E', and tan δ) of drawn (5X) nylon 66 fiber in D.I. water and tetrachloroethylene liquid media and in air at 0% RH. The reproducibility of the data is about ±3%. The α-peak temperature in tan δ of nylon 66 in water and tetrachloroethylene are shifted about 90 and 40°C lower than in air at 0% RH.

The occurrence of the α-transition (maximum in tan δ) at 110°C for the highly oriented nylon 66 fibers is typical of the values obtained by other workers for similarly prepared nylon 66 [12,13]. The effect of moisture on nylon 66 in terms of the α-dispersion has been reported by Woodward et al. [12]. They found the shift to lower temperature for nylon to be 15°C per percent of moisture.

(2) Effect of solvents on poly(ethylene terephthalate). The interaction of organic liquids with poly(ethylene terephthalate) (PET) during processing has received increasing attention in applications to dyeing and finishing. The adoption of organic liquids as processing media means that essentially hydrophobic synthetic fibers will be in intimate contact with hydrophobic liquids at elevated temperatures for extended periods of time. It is entirely possible that interaction of these organic liquids with fibers such as polyesters and polyamides may alter the "fine structure" (morphology) of these fibers more drastically than similar processing in water.

Tetrachloroethylene is the dominant nonaqueous medium under development for dyeing. The kinetic and thermodynamic treatments found to describe disperse dyeing of cellulose triacetate, nylon 6 and 66, and polyester from water have been successfully applied to disperse dyeing from tetrachloroethylene [14-18]. The dyeing of hydrophobic fibers by nonionic dyes in an aqueous solution is described by a model in which dye molecules diffuse across a boundary

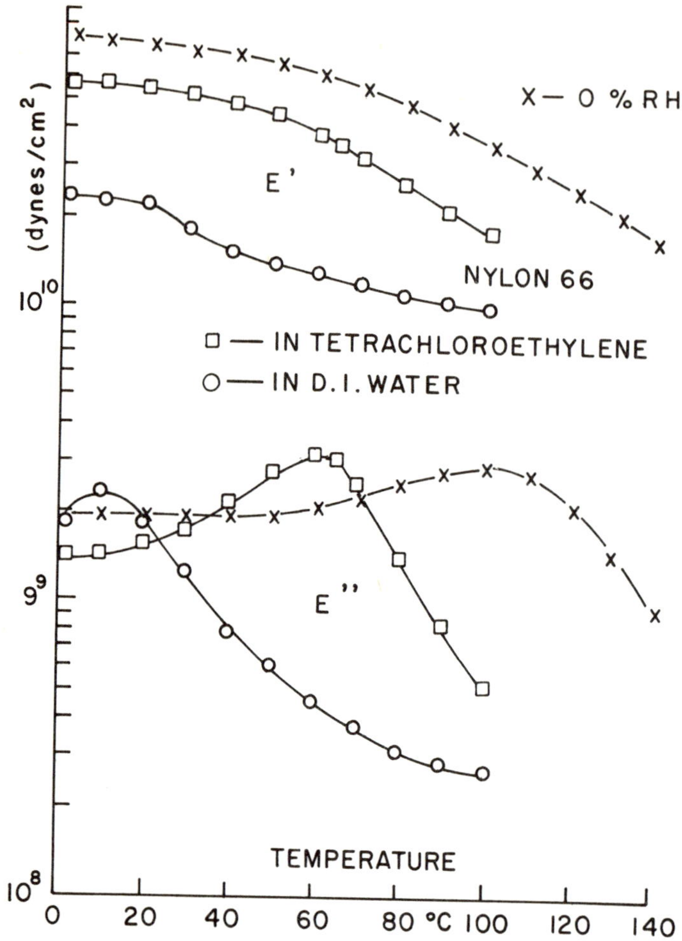

Fig. 7.8. Dynamic modulus (E') and loss modulus (E") of nylon 66 in D.I. water and tetrachloroethylene.

layer and are absorbed at the fiber surface and then proceed to diffuse into the polymer. From an organic medium, diffusion within the fiber appears to determine the overall rate of dyeing. This diffusi approximately follows Fick's laws of diffusion. Dyeing rates from tetrachloroethylene at a given temperature above 80°C are generally much larger than from water, reflecting the plasticization of the fiber by the organic liquid. Diffusion coefficients and activation energies for diffusion of disperse dyes exhibit marked dependence on the molecular sizes of the dyes, suggesting involvement of polymer segmental mobility and free volume in the penetration process.

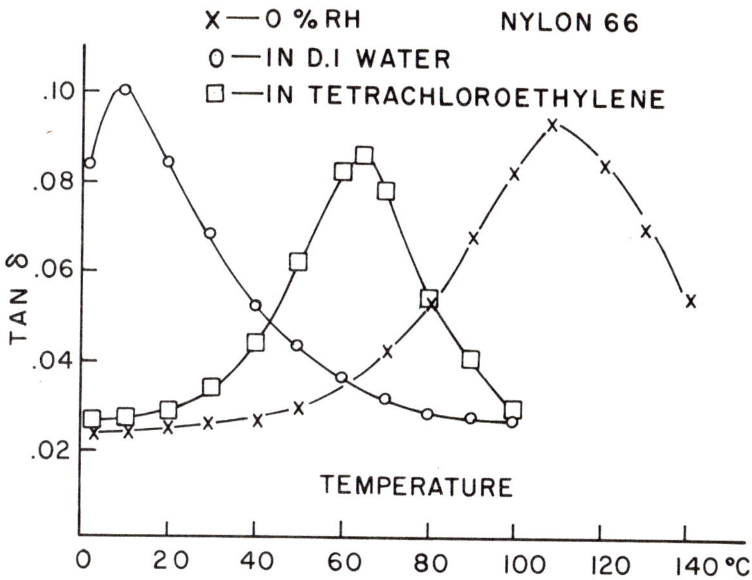

Fig. 7.9. Loss tangent (tan δ) of nylon 66 in D.I. water and tetrachloroethylene.

The rate of sorption and the equilibrium sorption of tetrachloroethylene also depends upon the morphology of the poly(ethylene terephthalate) fiber [19]. The sorption of perchloroethylene above 80°C by commercial fibers and films results in changes in the polymer morphology [20,21]. These structural reorganizations accompanying sorption have been demonstrated for a number of liquids [11,21,22]. Dynamic mechanical analysis of polymers in a liquid medium has become an important tool for studying structural changes in these solvents [23].

Figures 7.10 and 7.11 show the dynamic modulus (E'), loss modulus (E"), and tan δ of drawn polyester fibers (5X in deionized water, tetrachloroethylene, and air at 0% RH. The fibers were not preconditioned prior to the measurements. The α-dispersion peak of polyester in deionized water was not reached in this experiment, but the tan δ curve was shifted to the left of the curve for air at 0% RH. Dynamic mechanical data collected on the polyester filaments in a 100% RH atmosphere using the Rheovibron were nearly superimposable on the data collected for the filaments in water shown in Figs. 7.10 and 7.11. The α-dispersion peak for drawn polyester in tetrachloroethylene was shifted to 85°C from a value of 130°C in air at 0% RH.

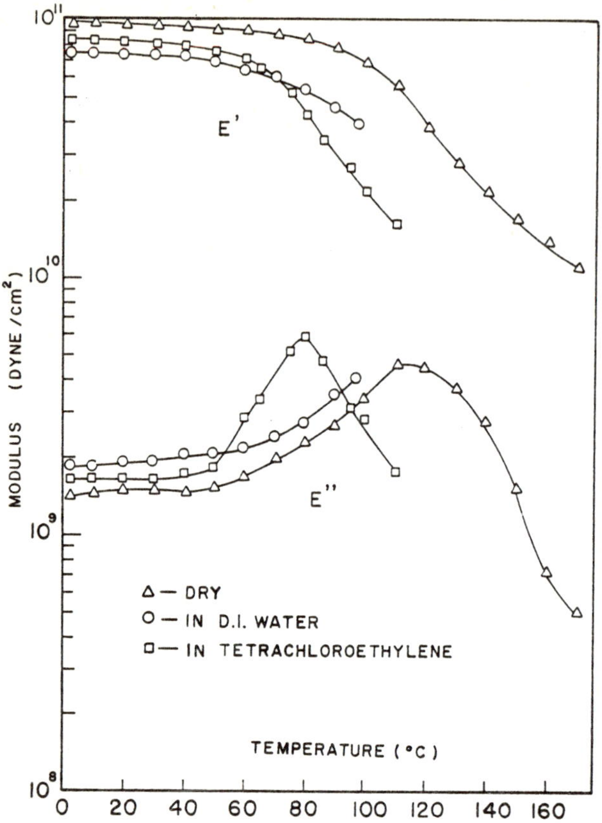

Fig. 7.10. Dynamic modulus (E') and loss modulus (E") of polyester in D.I. water and tetrachloroethylene.

The dynamic mechanical properties were measured at a 1°C/min heating rate. If diffusion of the liquid medium into or out of the fiber is slow at this heating rate, then the measured dynamic mechanical properties may actually be for filaments only partially penetrated by the liquid. Thus, the dynamic properties at a given temperature may exhibit a time dependence, reaching equilibrium values when the sorption of the liquid and any accompanying structural reorganization reach equilibrium. However, dynamic mechanical data indicating transitions under nonequilibrium sorption conditions may have practical relevance to batch dyeing where the fiber in the dye bath is heated at a programmed rate to the dyeing temperature and then dyed for a short time.

In Fig. 7.12 are presented the dynamic mechanical properties of fibers conditioned in tetrachloroethylene for 120 hours at 25°C

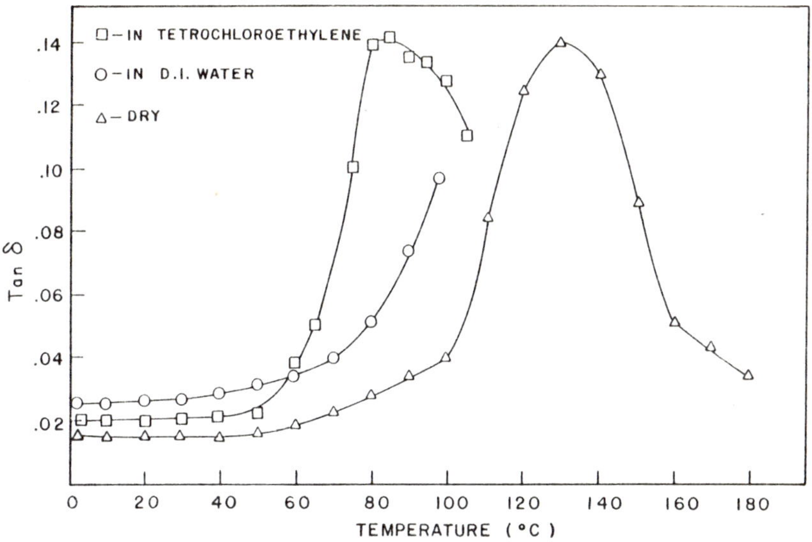

Fig. 7.11. Loss tangent (tan δ) of polyester in D.I. water and tetrachloroethylene.

before testing. A fiber which was free to contract is compared with one held at constant length during the conditioning. The sample was immersed in tetrachloroethylene during measurements. Sorption data indicate that an equilibrium uptake of 11 to 12% liquid should have been achieved during this conditioning [20]. During the course of the temperature increase in the Rheovibron experiment, there would be a desorption of liquid with increasing temperature, tending toward an equilibrium value of 8 to 9% at 120°C [20]. Fiber shrinkage also occurs as the temperature of measurement increases [24]. Diffusion rates indicate that these equilibrium values of desorption could be readily approached during the heating schedule of the Rheovibron experiment above 60°C. The maximum of the α-dispersion peak for fibers held for 120 hours in tetrachloroethylene is shifted to 110°C from 80°C for fibers not preconditioned. This shift is probably a result of constraints [25] placed on the amorphous regions by small crystallites induced [21] during the prolonged immersion. Knox et al. [21] have cited evidence for such morphological rearrangement during liquid immersion and proposed that the rearrangements are induced by annealing below 180°C. The values of the maximum in the loss modulus, E''_{max}, are decreased by preconditioning in tetrachloroethylene (compare Figs. 7.10 and 7.12). Also, the breadth of the

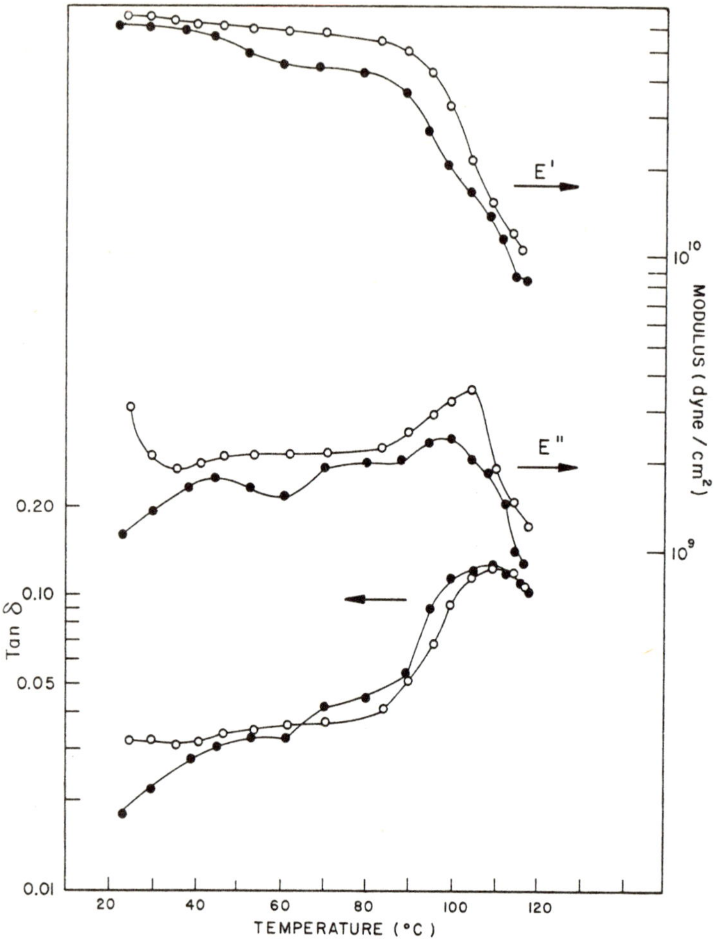

Fig. 7.12. Comparison of dynamic mechanical properties of fiber preconditioned in tetrachloroethylene at 25°C for 120 hr; open circles = no restraint, filled circles = constant length.

E'' peak is considerably increased by the preconditioning. Both changes in the E'' peak are consistent with increased crystallinity due to preconditioning [25]. The effects of crystallite size and distribution and orientational factors on the α-dispersion have been studied in detail for thermally annealed fibers [25]. Comparison of the fiber held at constant length during the conditioning with the unrestrained fiber shows that orientational changes have also occurred. The dynamic-modulus and tan δ curves for the restrained fiber differ considerably in the 50-to-100°C range from those observed for the unrestrained fiber. This difference probably reflects the relaxation of tensions in oriented amorphous regions in the restrained fibers which were relaxed during the immersion for the unrestrained fibers.

The tan δ, E', and E" values for unrestrained fiber conditioned in tetrachloroethylene at 110°C for 3 hours are presented in Fig. 7.13.

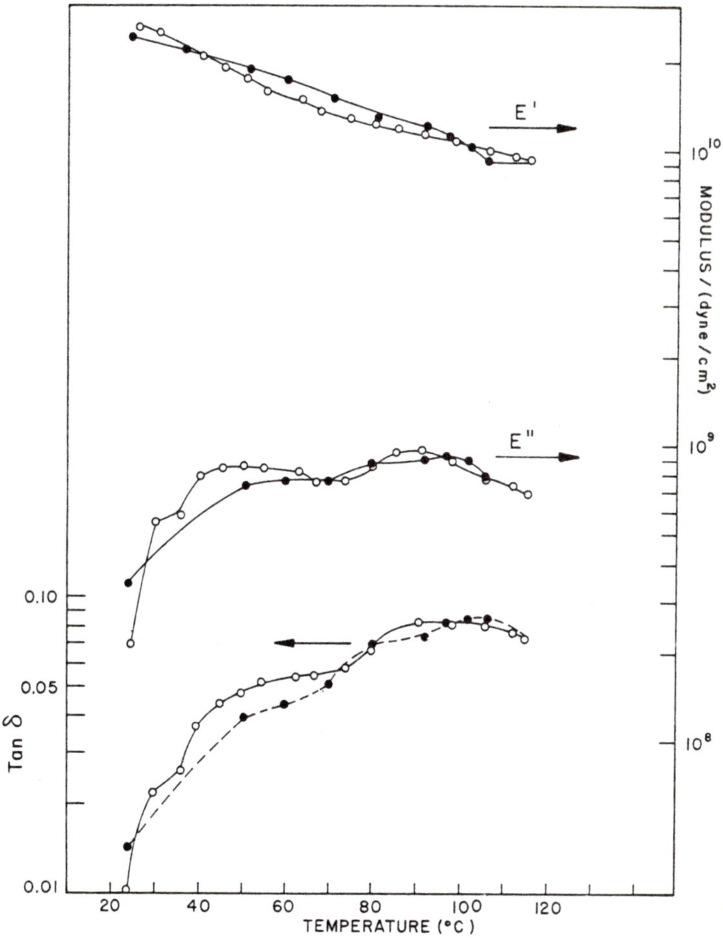

Fig. 7.13. Dynamic mechanical properties of fiber preconditioned in tetrachloroethylene at 110°C for 3 hr; open circles = increasing temperature; filled circles = decreasing temperature.

This conditioning allowed attainment of equilibrium sorption of 9% liquid, and since it took place at or above the glass temperature of the fiber in tetrachloroethylene, considerable morphological rearrangement occurred. The sample was held at room temperature in liquid for 16 hours prior to the Rheovibron experiment. The E" curve is peculiar in shape, exhibiting two humps, one centered at 50°C and one at 90°C. The maximum of the tan δ curve is located at 95°C, and the maximum value is lower than that shown in Fig. 7.11 or 7.12. Since the E" peak may be associated with the energy dissipated hysteretically as heat in the amorphous regions of the polymer in the T_g transition region, the broadening of the peak is evidence for a morphological rearrangement resulting in a highly plasticized amorphous region. Such broadening is often cited as evidence for an increased range of order.

During the Rheovibron experiment the sample was first heated to 115°C and then cooled to room temperature at the programmed rate. The dynamic data at a given temperature were nearly superimposable, whether that temperature was approached from room temperature or from 115°C. Such reversibility in the α-transition region indicates that no further morphological rearrangement is occurring during the time interval of the Rheovibron experiment, and that the heating is slow enough for absorption and desorption toward equilibrium values at a given temperature not to affect the dynamic data.

Dynamic mechanical data for an unconditioned commercial biaxially oriented polyester film are shown in Fig. 7.14. These data will be used for correlation with data taken from the literature on the non-aqueous dyeing rate of films. The Rheovibron data were obtained by heating the sample from room temperature to 116°C and then cooling back to room temperature. The data for increasing and for decreasing temperature are not superimposable. This may reflect morphological changes induced during the liquid immersion and/or the nonattainment of equilibrium sorption of tetrachloroethylene at lower temperatures. The disagreement is more pronounced at lower temperatures than at higher temperatures.

The drastic changes that are induced in oriented fibers by immersion in tetrachloroethane have been illustrated by Lawton and Murayama [23]. Fibers were conditioned for 120 hours at 25°C in tetrachloroethane with and without restraint. Approximately 19% shrinkage occurred in the unretrained sample. Knox et al. [22] have shown that such a treatment would induce appreciable increases in crystallinity. The dynamic data are clearly shifted by prolonged exposure to the liquid or by tension applied to the fibers preventing shrinkage and disorientation. In Fig. 7.15, the extreme plasticizing action of the liquid is evidenced by the dynamic data obtained

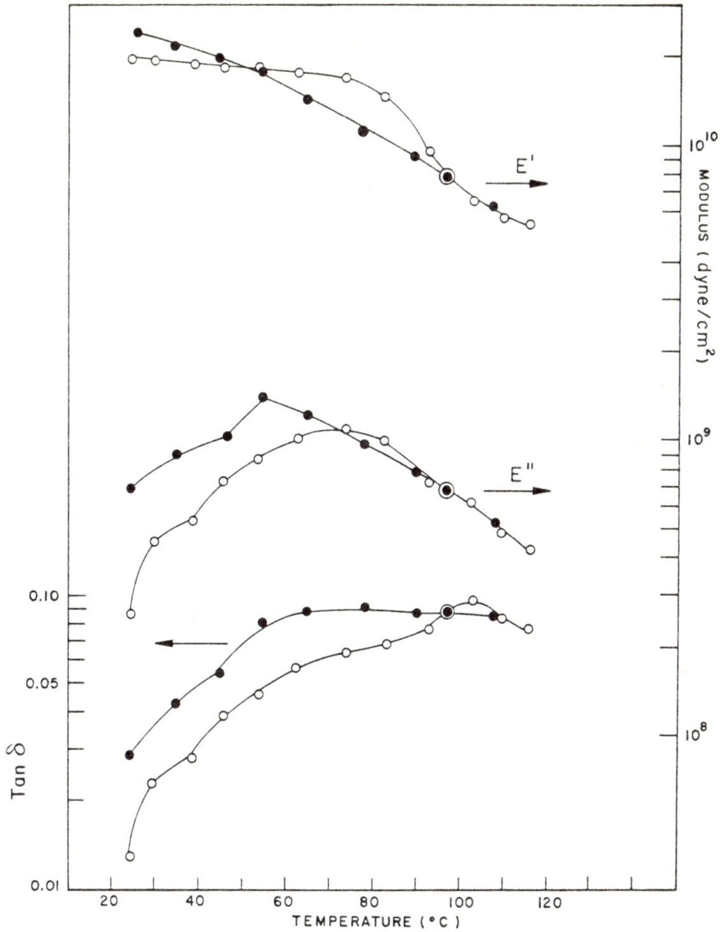

Fig. 7.14. Dynamic mechanical properties of film in tetrachloroethylene; open circles = increasing temperature; filled circles = decreasing temperature.

upon cooling the sample from 102°C to 25°C. This sample had shrunk 26% while being heated from 25°C to 102°C in a relaxed condition in the Rheovibron experiment.

The expected lowering of E" and tan δ as the sample was cooled was not observed. Apparently the sample underwent such drastic rearrangement at the elevated temperatures during the Rheovibron experiment that it remained in a rubbery state when cooled. Ribnick and Weigmann [21] have noted that commercial fibers lose their fibrous structure (dissolve) above 120°C in tetrachloroethane. Since the fibers were conditioned for an extended period at 25°C in

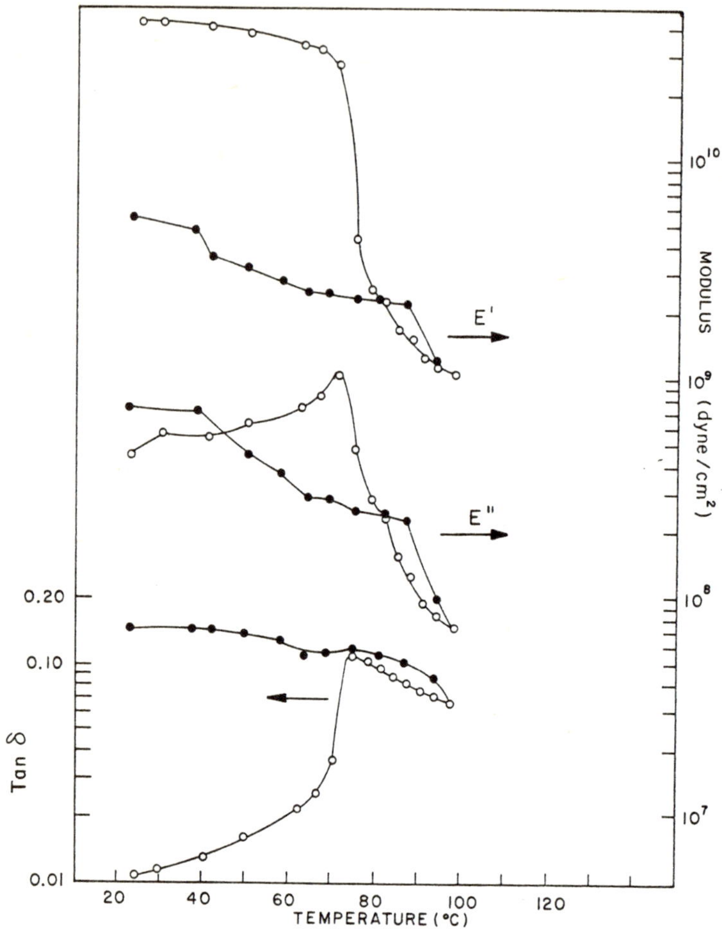

Fig. 7.15. Dynamic mechanical properties of fiber preconditioned at constant length for 120 hr at 25°C in 1, 1, 2, 2-tetrachloroethane; open circles = increasing temperature; filled circles = decreasing temperature.

tetrachloroethane prior to the Rheovibron experiment, it appears unlikely that the increased plasticization upon cooling is due to an increased penetration of the fiber at the higher temperatures. It might be speculated that the extreme interaction (as evidenced by crystallite dissolution above 120°C) of tetrachloroethane with the fiber results in a plasticized rearranged structure that remains rubbery below room temperature over the time interval of the Rheovibron experiment. This state of the fiber might be related to the formation of sol-gel structures in concentrated polymer solutions

[26]. The crystallites in the polymer would essentially melt at 110°C in tetrachloroethane and re-form small crystallites as the temperature was lowered, resulting in an elastic nonflowing gel. The role of loss of polymer and oligomer from the fiber due to dissolution in the liquid also needs clarification. Tetrachloroethane [27] and other liquids [11,20,28] leach considerable material from the fiber at elevated temperatures.

The structural alterations due to immersion in tetrachloroethane and the powerful interaction of the polymer with tetrachloroethane are illustrated by the following comparisons. For the unconditioned fiber over the temperature interval of 70 to 94°C, E" decreases from 4.4×10^9 to 7.2×10^8 dyn/cm^2, while E' decreases from 4.2×10^{10} to 1.9×10^{10} dyn/cm^2. For the unrestrained fiber preconditioned at 25°C for 120 hours, from 80 to 103°C, E" decreases from 2.0×10^9 to 1.7×10^7 dyn/cm^2 and E' from 5.3×10^{10} to 4.2×10^8 dyn/cm^2. The fiber preconditioned at constant length also exhibits large decreases in E" and E' over the temperature interval from 70 to 95°C: E" from 1.0×10^9 to 8.0×10^7 dyn/cm^2, and E' from 2.8×10^{10} to 1.1×10^9 dyn/cm^2. These large decreases in E" are further evidence for the solvated (rubbery) state of the fiber in tetrachloroethane at temperatures approaching 100°C. For fiber preconditioned at constant length in the less interacting tetrachloroethylene, E" would typically decrease from 2.0×10^9 dyn/cm^2 at 70°C to 1.9×10^9 dyn/cm^2 at 110°C, and E' from 4.6×10^{10} to 1.4×10^{10} dyn/cm^2.

Changes in the dynamic mechanical properties of polymers after immersion in a liquid may reflect alteration of viscoelastic properties of the polymer due to interaction with the liquid and also alteration of the polymer structure due to changes in crystallinity and orientation induced by the interaction. The degree of disorientation of fibers occurring during the heating interval of the Rheovibron experiment is indicated in Table 7.4 by the shrinkages of fibers with and without preconditioning.

Murai and Kuroda [8] have reported measurements of dynamic mechanical properties of polyester immersed in liquids. Their measurements were on amorphous films immersed in aqueous phenol solutions, using the Iwamoto Manufacturing Company viscoelastic spectrometer. They found large depressions of the polymer glass temperature which depended upon the concentration of phenol in the fiber. They also measured changes in E' and E" at fixed temperatures during sorption of phenol into the film. Murai and Kuroda have presented dynamic mechanical data indicating that for nonequilibrium

TABLE 7.4

Shrinkage and resonance peak data for sample while being heated in liquid during the Vibron experiment

Polyester preconditioning in immersion medium	Immersion medium	Temp. interval, °C	Shrinkage in Vibron, %	Temp. at which max. occurs in E" peak, °C	Max. E", dyn/cm²	Temp. at which max. occurs in tan δ peak, °C	Max. tan δ (E"/E')
None	Dry air	0-160	6.1	110	4.8×10^9	130	0.140
None	Water	0-98	1.1	--	--	--	--
None	Tetrachloroethylene	0-110	6.3	80	5.8×10^9	85	0.142
120 hr at 25°C, unrestrained	Tetrachloroethylene	25-117	8.6	105	3.6×10^9	110	0.125
120 hr at 25°C, constant length	Tetrachloroethylene	25-118	9.3	98	2.5×10^9	110	0.130
3 hr at 110°C, unrestrained	Tetrachloroethylene	25-115	0	91[a]	9.6×10^8[a]	91	0.084
18 hr at 110°C, unrestrained	Tetrachloroethylene	25-115	0	70 to 85[a]	1.1×10^9[a]	85	0.086
None	Tetrachloroethylene	25-115	0	74	1.1×10^9	103	0.096
3 hr at 85°C, unrestrained	Trichloroethylene	25-83	0	62[a]	1.4×10^9[a]	81[a]	0.086[a]
None	1,1,2,2-tetrachloroethane	0-94	38.2	67	4.4×10^9	73	0.098
120 hr at 25°C, unrestrained	1,1,2,2-tetrachloroethane	25-98	31.4	95	2.1×10^9	95	0.074
120 hr at 25°C, constant length	1,1,2,2-	25-102	25.9	71	1.0×10^9	75	0.110

[a]Very broad and ill-defined peak.

sorption of phenol both the glass transition of the regions of the polymer penetrated by the liquid and that of the unpenetrated polymer can be detected.

Zachmann and Scherman [9] have measured dynamic mechanical properties of polyester films swollen with organic liquids under several conditions. The glass temperatures of the swollen films were markedly depressed. Similar experiments were performed by Illers and Breur [29] in their pioneering work on dynamic mechanical properties of polyesters.

4. EFFECTS OF HUMIDITY

Dynamic mechanical properties of polymers are affected by humidity. The water can be considered as a plasticizer for polar polymers. The T_g of nylons and polyvinyl alcohol is greatly reduced by water [12, 30-35]. A number of dynamic mechanical analyzers are equipped with additional systems for controlling the relative humidity in the sample chamber. Woodward et al. [12] studied the damping factor ($\tan \delta$) and the dynamic modulus for nylon 66 at various humidities as a function of temperature.

The loss-tangent peak of the primary dispersion region of nylon 66 shifts towards a lower temperature with increasing moisture content. The large decrease in the dynamic modulus as a function of temperature is observed with increasing moisture content. They conclude that in polyamides the α-peak (glass transition region) is associated with segmental motion in amorphous regions as a consequence of the breaking up of hydrogen bonds (or polar bonds) between chains.

The α-peak temperature T_g is shifted downward with increasing water content in nylon 66: 97°C for 0%, 72°C for 1.02%, 37°C for 3.7%, and 2°C for 7.21% water content by volume. However, the intensity of the α-peak ($\tan \delta_{max}$) did not change a great deal as a function of water content. The significant result is that the shift of the α-dispersion is about 15.0°C per percent moisture content in the case of nylon 66.

A similar shifting of the α-peak by water is presented in Fig. 7.16. This shows the temperature dependence of the dynamic moduli of a polyvinyl alcohol (PVA) sample which has been made to absorb water up to 8%, 20%, and 30%, and of the dry sample [35]. The α-dispersion, located in the neighborhood of 75°C in the dry sample, shifts conspicuously to lower temperatures with the increasing moisture absorption, the values being 15°C, -8°C, and -20°C. This is due to the absorbed moisture acting as a plasticizer in the non-crystalline region.

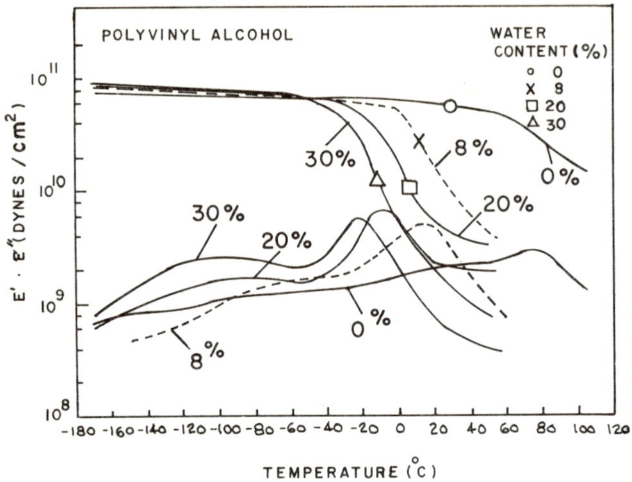

Fig. 7.16. Effect of hydration on the dispersion curves of polyvinyl alcohol at 138 Hz (after Takayanagi, 35).

The dynamic loss curves of PVA samples show very weak broad dispersion in the temperature range from -60°C to -180°C. The intensity of this low-temperature peak increases with increasing the moisture absorption. This dispersion is characteristic of moisture absorption, which behaves similarly in the hydration dispersion in polyamide in the neighborhood of -50°C. With PVA, when the amount of water absorption reaches 20-30%, the position of the hydration dispersion shifts markedly downward in temperature. With polyamide, however, such a shift is not clearly observed, since not much moisture absorption actually takes place. As for the mechanism of hydration dispersion in polyvinyl alcohol, it is presumed to be due to the local twisting mode of the main chains under conditions where water molecules have made hydrogen bonds with hydroxyl groups of main chains. An increase in intensity with moisture absorption is due to the occurrence of an increased hydration state due to severing of the intermolecular hydrogen bonds by water molecules. Water affects the low-temperature (secondary) transition in many cases, and it also often is the cause of a new secondary transition in the region from -50 to -100°C [36-42].

5. DIFFUSION

The dynamic parameters (E', E", and tan δ) of polymers in liquids give useful information not only on mechanical properties of polymers but also on diffusion.

Fujita, Kishimoto, and Matsumoto [43] have shown that above the glass transition the diffusion coefficient D of organic liquids in an amorphous polymer, polymethyl acrylate, is related to the steady-state viscosity η by the equation

$$\ln \frac{D_T}{RT} = - B \ln \eta + C, \qquad (7.16)$$

where B and C are constants; their experimental results indicate that B is near unity.

The shear viscosity η may be replaced by E''/ω, where E'' is the tensile dynamic loss modulus measured at frequency ω and at temperature T [44]. Since E'' is the tensile loss modulus, the quantity E''/ω is equivalent to the tensile viscosity η_T, which is equal to 3η. Substitution of E''/ω in equation (7.16) leads to

$$\ln \frac{D_T}{RT} = - B \ln \frac{E''}{\omega} + \left[C - B \ln \frac{1}{3} \right]. \qquad (7.17)$$

Strictly speaking, this replacement is rigorous only for E'' measured at zero frequency. The validity of the substitution is further discussed in [45].

Equation (7.16) with E''/ω substituted for η was applied to the diffusion of acid dyes in drawn nylon 66 fibers by Bell [46]. It was assumed that a measurement frequency of 11 Hz was low enough for the substitution to be valid.

The relation

$$\ln \frac{D_T}{RT} = - 4.5 \ln \frac{E''}{\omega} + C' \qquad (7.18)$$

was found, where E'' was measured under dyeing conditions. A different value of C' was required for each draw ratio, although the C' values were independent of temperature. The fact that the value of B was 4.5 instead of near 1 was attributed to the fact that the polymer was oriented (E'' was necessarily measured parallel to the fiber axis, whereas diffusion takes place primarily perpendicular to the fiber axis) and semicrystalline, and also possibly to the above substitution of E''/ω for η.

The relationships between dye diffusion and the dynamic loss modulus under the dyeing conditions have been observed in polyester fiber [23,45], acrylic fiber [47], and polypropylene film [48]. These indicate that the diffusion is controlled by the mobility of the polymer chain segments.

Fujita et al. [43] originally derived their equations in a manner similar to the phenomenological Williams-Landel-Ferry (WLF) equation [50] of linear viscoelasticity, both being based upon the free-volume equation for the viscosity of simple liquids established by Doolittle [51]. Fujita [52] later derived his equation from the theory of Cohen and Turnbull [53].

Kojima and Iijima [49] have presented experimental evidence of the applicability of Fujita's free-volume theory to the diffusion of disperse dyes in semicrystalline poly(ethylene terephthalate). Rosenbaum [54] suggested that variations in diffusion coefficients of dye into acrylic fibers with temperature could be represented by an equation having the WLF form. Gur-Arieh and coworkers [55-57] have gathered extensive evidence for acrylics and polyester indicating that diffusion behavior above T_g can be represented in terms of the WLF equation. They stress that the dyeing process is controlled by the segmental mobility of the polymer. Their studies indicate that dye carriers depress T_g and increase this segmental mobility. Kitamura [58] has also demonstrated that temperature dependence of diffusion coefficients for disperse dyeings of polyester from water or tetrachloroethylene may be expressed in terms of the WLF equation.

In this section the experimental studies on the relation between dynamic mechanical properties and dye diffusion behavior in polyester and acrylic fibers will be discussed in some detail.

(1) Poly(ethylene terephthalate) (PET). Extensive studies on the effect of structural changes on dye diffusion in poly(ethylene terephthalate) have been carried out using a wide variety of crystallinities and orientation [3]. The starting material in this study was unoriented PET yarn (birefringence 0.004). Yarns with draw ratios of 3X and 4.25X were obtained by drawing the spun yarn over a hot pin at 80°C.

Samples with different degrees of heat setting were obtained by freely suspending skeins of yarn in an evacuated oven for periods of 6 hours. It had been previously determined that no further changes in crystallinity occurred after 6 hours, so that the structures could be assumed dependent only on the temperature of heat setting and not the time. This heat-setting treatment was followed by a 1-hour boil in water, during which the samples were again free to relax. The water treatment was given so that no structural changes due to water would take place during the dyeing experiments. X-ray crystallinities were found unchanged for longer times of boiling after the initial hour. The structural measurements were then made before dyeing.

X-ray measurements were made after winding the yarn samples on aluminum frames. Crystallinities were measured on a Norelco diffractometer using the method of Dumbleton and Bowles [59]. Crystallite orientations were obtained from azimuthal diffractometer scans on the ($\bar{1}05$) plane. Crystal-size determinations perpendicular to the fiber axis (for drawn samples) were obtained from 2θ scans on the (100) plane. The instrumental broadening was obtained from the width of the diffraction peaks of a sample of hexamethylene tetramine. The Gaussian form of the instrumental broadening correction was used [60]. Small-angle measurements were made on a Kratky camera with slit collimation and on a small-angle pinhole camera. The small-angle data were used to obtain crystal sizes along the chain axis for the drawn samples [61]. The orientation of the fibers was also judged from birefringence measurements which, when combined with the x-ray data, yielded the orientation in the noncrystalline regions [62].

Dynamic measurements were made on a Rheovibron viscoelastometer which was operated at a frequency of 11 Hz. Measurements were made with the sample in a nitrogen atmosphere at 0% RH and at 100% RH. The modulus E' and the loss factor tan δ were measured at 5°C intervals, with smaller intervals near the transition region. Samples were allowed to equilibrate for 5 minutes at each temperature before taking the measurements.

Dyeing-rate and saturation measurements were made without carriers or other dye-bath additives, so that dyeing rates were relatively low. No difference in dyeing rate or dye absorbed at saturation was found between samples which had been boiled for 1 hour in distilled water and samples which were given a scour after the boiloff; it was concluded that the additional scour was unnecessary, and it was discontinued. Before dyeing, the fiber samples were conditioned at 21°C and 65% RH.

The dye used was purified Eastman C.I. Disperse Red 15, 1-amino-4-hydroxyanthraquinone. The dye bath was prepared by adding 4.5 g of the purified dye to 3 liters of distilled water. A fiber sample, weight 0.020 g, was placed in a 250-ml bottle together with 120 ml of the dye solution. The bottle was quickly capped and submerged in an oil bath at the dyeing temperature. The dyeings were made at 100°C, with the exception of a few measurements at 80°C on undrawn yarn.

Dyeing times of from 25 minutes to 70 hours were used. These times do not include the 4 minutes required for the dye bath to reach temperature. At the end of the dyeing the bottles were removed from the oil bath and cooled in tap water. The fiber sample was removed from the bottle and washed, first in water and then in acetone, to remove surface dye. For the acetone wash, the fiber was placed in a sample bottle with acetone and shaken vigorously. This rinsing was carried out five times, after which the fiber was dried, conditioned at 21°C and 65% RH, and reweighed. The dye was extracted from the fibers with hot dimethylformamide at 95°C. Five extractions were made on each sample, the first three being for 5 minutes and the last two for 15 minutes. The extract was brought to 100 ml with dimethylformamide, and the optical density of the extract was measured with a Cary Model 11 spectrophotometer. The percentage of dye on the fiber was calculated from the following formula:

$$\%\text{ dye absorbed} = \frac{(\text{optical density}) \times (\text{extract volume}) \times (\text{dilution}) \times 10}{(\text{extinction coefficient}) \times (\text{dyed-sample weight}) \times 1000}$$

Since there was more than enough dye to saturate the dye bath under the conditions used, the dyeings can be considered as occurring in an "infinite" dye bath in which the dye concentration does not change. The dye concentration for the 80°C dyeings was lower because of the lower solubility at that temperature. This is compensated by the lower saturation value, C_∞, used in the diffusion-constant calculations

Diffusivities were calculated from graphs of dye on fiber vs. time by the method of Hill [63]. Theory states that when the sample is not near dye saturation there should be a linear relation between dye on the fiber and the square root of time. In Figs. 7.17 and 7.18 data for the undrawn yarn drawn 4.25X are plotted. It is seen that the graphs are linear.

X-ray results for the conditioned, heat-crystallized, spun PET yarn indicate that the crystallinities increase with an increase in annealing temperature, although the range of crystallinities is small when compared with the range of crystallinities obtained for heat-crystallized, spun PET samples which have not been boiled in water [25]. A 1-hour boil in water for an as-spun PET yarn produces a crystallinity of 37%. To obtain this crystallinity using dry heat only requires an annealing temperature of approximately 125°C. Thus immersion in water appears to be equivalent to an increase in annealing temperature of about 25°C. This is close to the decrease in the glass transition temperature of undrawn PET by water [29,64], which implies that the crystallization temperature is fixed relative to the glass transition temperature. The ultimate attainable crystallinity does not seem to be affected by the boiling-water treatment.

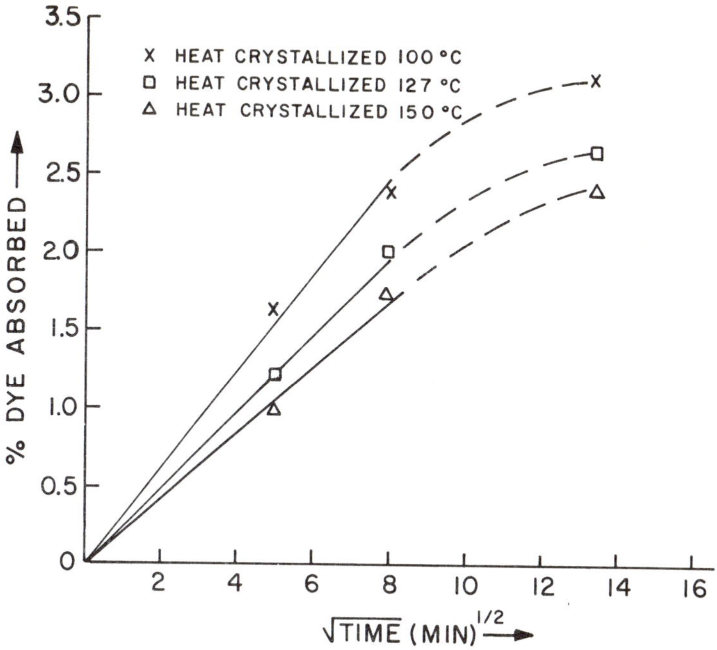

Fig. 7.17. Percent dye on fiber versus $(\text{time})^{1/2}$ for undrawn fibers.

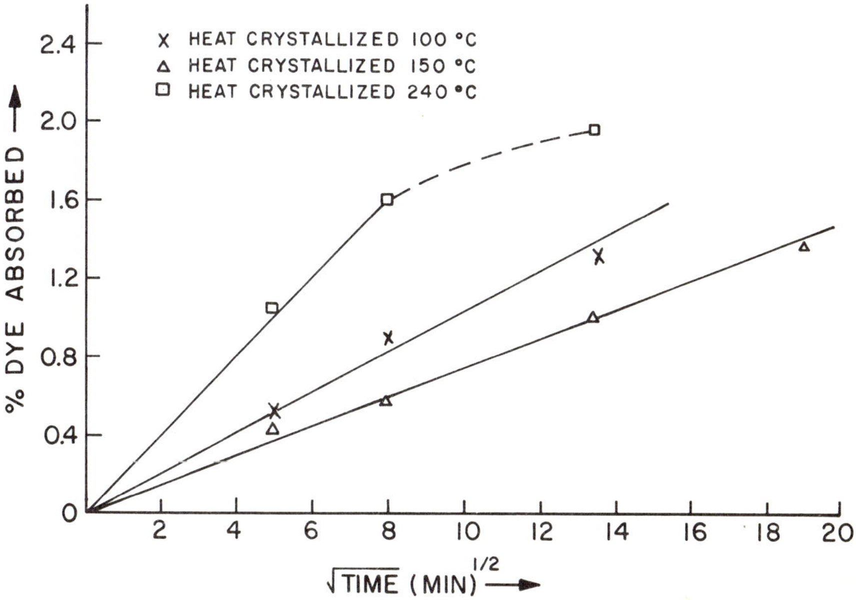

Fig. 7.18. Percent dye on fiber versus $(\text{time})^{1/2}$ for fiber drawn 4.25X.

The long period increases with annealing temperatures, but is lower than the corresponding values before the boiling-water treatment. The intensity of the small-angle maximum is also much lower after the water treatment. This indicates that boiling does not simply increase the crystallinity but causes a decrease in the crystal size. Possibly a large range of crystal sizes is present, since this could account for the low intensity of the small-angle maximum.

An investigation of undrawn, heat-crystallized PET has shown [25] that the position of the α-transition (glass transition) depends on crystallinity, crystal size, and, in particular, the number of crystals, which may be represented by X/v, where X is the crystallinity and v the volume of a crystal. Figure 7.19 shows the graph of X/v

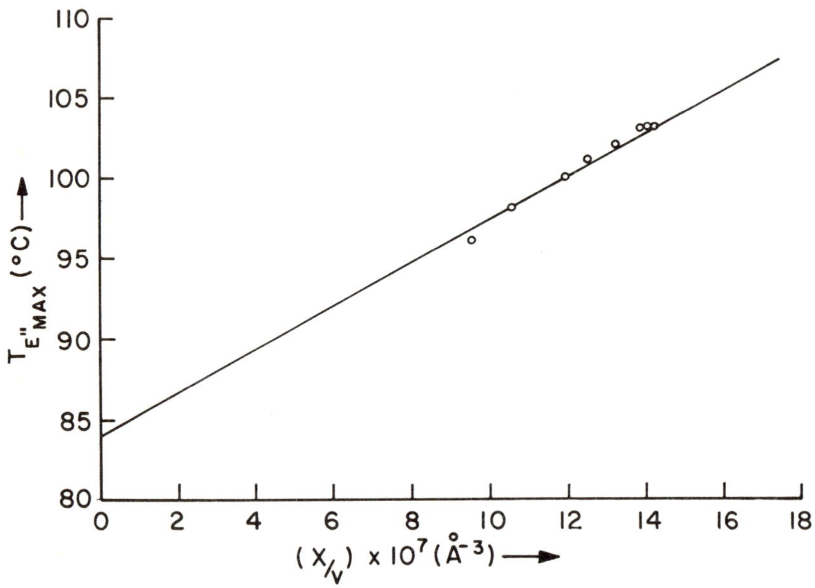

Fig. 7.19. $T_{E''max}$ versus X/v for undrawn, heat-crystallized PET fibers.

versus glass transition temperature. The points fit the line, which indicates the glass transition temperature is determined by the number of crystallites.

The dyeing results are shown in Figs. 7.20 and 7.21. The diffusivit D is plotted versus annealing temperature in Fig. 7.20; the equilibrium dye uptake C_∞, in Fig. 7.21. Both D and C_∞ exhibit a minimum at an annealing temperature near 175°C.

The diffusivity and dye uptake are much higher for the 100°C dyeing than for the 80°C dyeing. The reason for the much lower D values

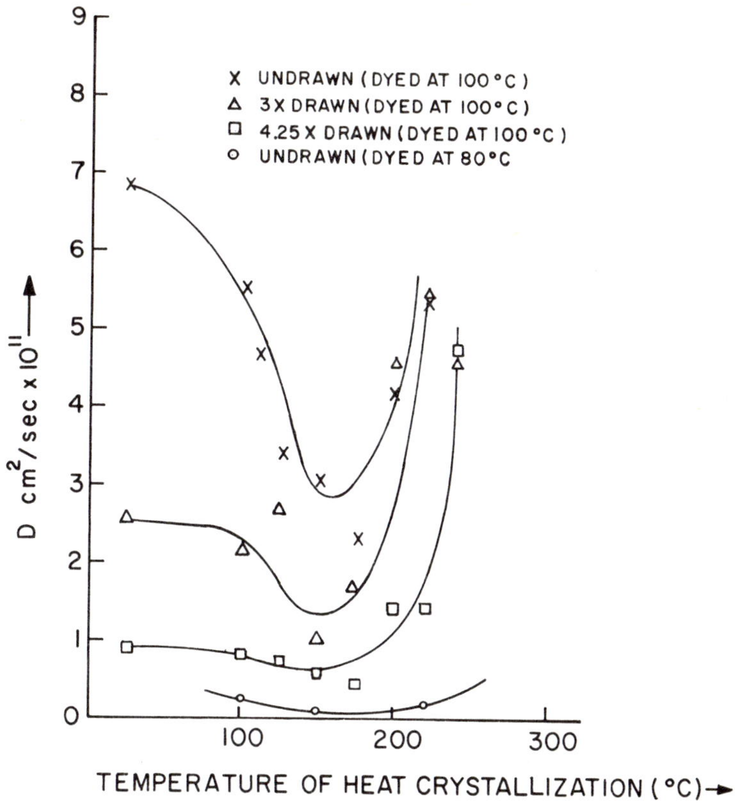

Fig. 7.20. D versus annealing temperature.

obtained for 80°C dyeing is evident from the E"-versus-temperature curves [3]. It is seen that 80°C is the glass transition temperature and corresponds to a maximum value in E". The E" curves fall quite steeply above 80°C, leading to a lower value of E" at 100°C and hence a high value of D. Alternatively, it may be said that at 100°C the chain segments have relatively large freedom of movement and dye molecules will easily penetrate the polymer, whereas at 80°C the polymer is almost in the glass region and molecular motion is much more difficult.

Values of E" at 100% RH and 100°C were taken from E" curves [3] by extrapolation to 100°C. The presence of water decreases the glass transition by some 20°C, but the height of the E" maximum is not appreciably altered. Since the height of E" is often taken as a measure of the crystallinity, this constancy implies that the crystallinity of the boiled and conditioned samples does not change on immersion in water.

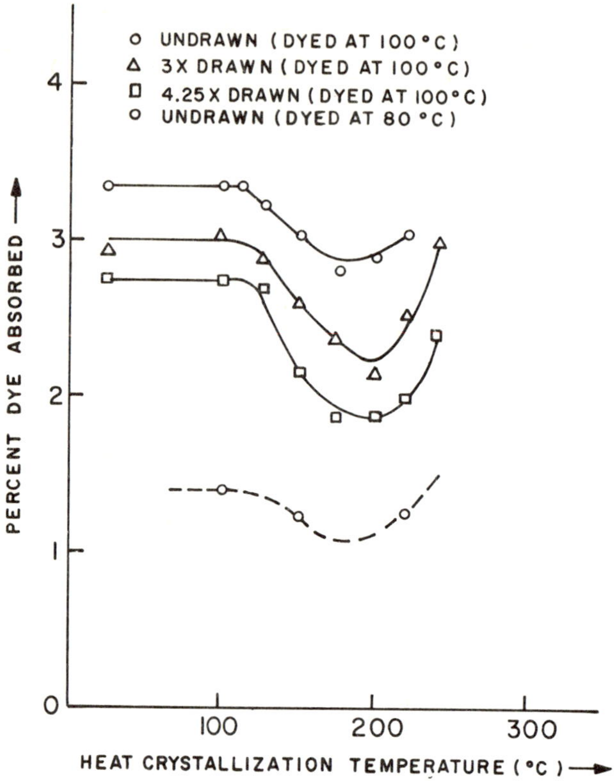

Fig. 7.21. C_∞ versus annealing temperature.

As PET absorbs approximately 2% water, the change in glass transition is some 10°C (% water), a value which is not very different from the 15°C/% found in nylon 66 [65]. This is perhaps surprising, since the molecules in PET are bound to each other by van der Waals forces, while in nylon 66 hydrogen bonds play a large part in intermolecular bonding. In fact the great difference between these two polymers is in the amount of water which is absorbed. Water is just as effective in lowering the T_g of PET as of nylon 66, but PET absorbs much less water than nylon 66 does.

Figure 7.22 shows log (D/RT), the molecular mobility, plotted against log E". A linear relationship is seen with the slope of the line near unity, as found by Fujita, Kishimoto, and Matsumoto [43].

The variation of E" with temperature at 100% RH gives a good indication of how a PET fiber will dye. The diffusivity depends on

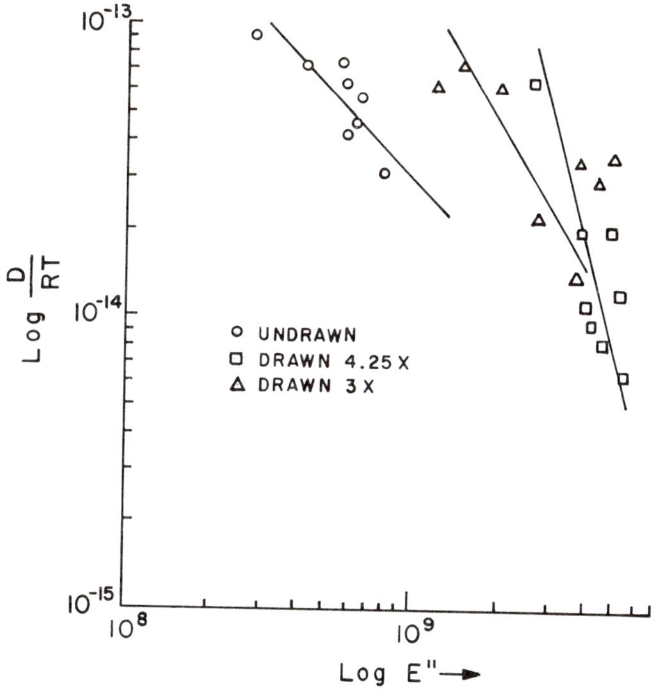

Fig. 7.22. Log D versus log E" (solid point denote T_g > 100°C).

the difference between the dyeing temperature and the glass transition temperature T_g. For a fixed dyeing temperature, D is dependent on the structure of the fiber, since T_g depends on orientation, crystallinity, and other structural parameters [25]. Usually, an increase in orientation or crystallinity will increase T_g. T_g may be decreased by chemical modification of the polymer or by the use of carriers (swelling agents) in the dye bath, and such a change can be considered equivalent to an increase in the dyeing temperature. It should be pointed out that it does not necessarily follow that a continual increase in the dyeing temperature will always continually increase D. The temperature at which PET will crystallize is related to the glass transition temperature, and if the temperature of dyeing is above the crystallization temperature T_c, then crystallization will occur during dyeing, which can decrease D by increasing T_g. As an example, consider undrawn, amorphous PET, which has a T_g of 80°C, measured at 0% RH on the Rheovibron, and a T_c of 98°C. The presence of water lowers T_g and T_c by some 20°C, and so dyeing at 100°C will cause crystallization. The T_g after crystallization is near 100°C (again as measured at 0% RH on the Rheovibron), and so the D will be lower than if the dyeing had been carried out at a somewhat lower temperature than that at which crystallization occurred. In drawn yarns the T_g is greater than 100°C (e.g., the T_g of the 4.25X yarn is 130°C at 0% RH), and so crystallization will not occur at 100°C in water. However, dyeing must be carried out under pressure or T_g must be lowered in order to get reasonable diffusivity, and so the problem of crystallization again arises.

The observation that D is sensitive to T_g immediately implies that segmental motion in the polymer is the limiting factor in determining the dye diffusion. Presumably the dye molecules will readily penetrate the polymer when the chain segments move so as to produce spaces equal to the size of a dye molecule. It has been shown [50] that the glass transition temperature occurs when segmental motion is such that spaces equal in size to a chain segment occur. Since this size is roughly equal to that of a dye molecule, the diffusivity would be expected to increase at the glass transition. Naturally, the larger the dye molecule, the higher must be the dyeing temperature to enable spaces of sufficient size to be produced by segmental motion.

The position of T_g is not the only factor in determining D, since D depends on the value of E" at the dyeing temperature and on the shape of the E" versus temperature relation, which will change with changes in structure.

The theory relating E" to D was originally proposed for diffusion in amorphous polymers above T_g. It was found that for drawn nylon 66 the relation $\log (D/RT) = -B \log (E''/\omega) + C'$ was valid, but B = 4.5 instead of the value of unity predicted by the theory. There was the possibility that $B \neq 1$ because a semicrystalline drawn material had been used or because the approximation $\eta = E/\omega$ was made. Since samples of widely differing structure were used in the present studies on PET, further insight is gained into the influence of structure on B. It is found that B = 1 for undrawn, semicrystalline PET (using $\eta = E''/\omega$), so that it appears that the fact that $B \neq 1$ in nylon 66 was due to orientation. Indeed, for the drawn PET samples B is greater than 1, with a larger value of B for the 4.25X yarn than for the 3X yarn.

In addition to the studies on the effects of structural changes on dye diffusion in PET, similar studies have been carried out on solvent dyeing. The diffusion coefficients of C.I. disperse Red 13 diffusing from perchloroethylene into fibers preconditioned for 3 hours at 110°C in the liquid have been obtained, as well as dynamic mechanical properties of this preconditioned fiber in tetrachloroethylene [23]. This result indicates a linear relationship between log (D/RT) and log (E"/ω) as predicted by equation (7.17). The diffusion in PET-tetrachloroethylene can be related to the loss modulus of PET in tetrachloroethylene. Values of diffusion coefficients of disperse dyes for diffusion from the tetrachloroethylene bath [14,18] and loss-modulus values of PET in tetrachloroethylene (Fig. 7.14) are used to construct the relationships between log (D/RT) and log (E"/ω).

Figures 7.23 and 7.24 present this relation for ten disperse dyes. A linear relationship is indicated. As the dyes increase in molecular

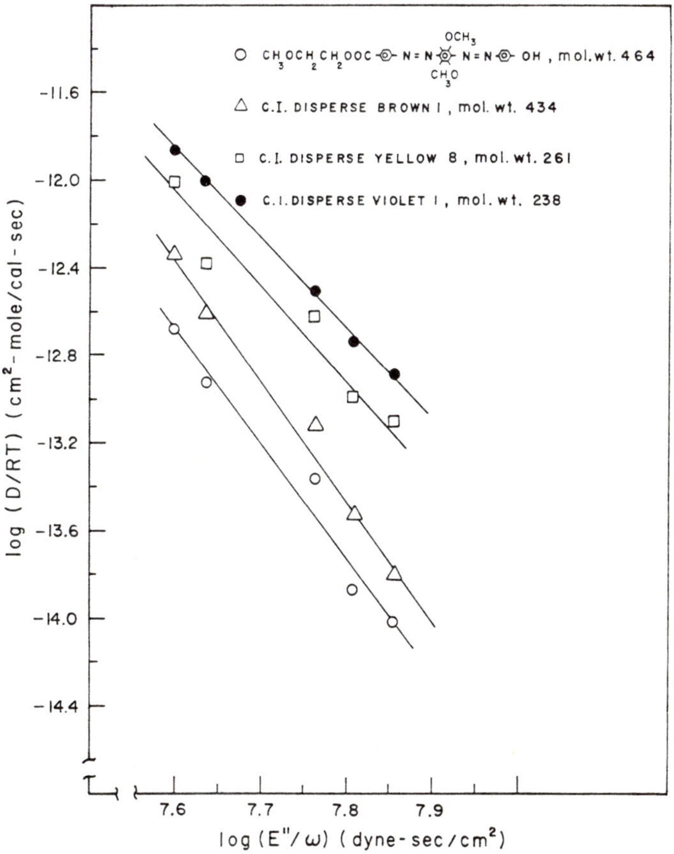

Fig. 7.23. Relationship between molecular mobility (D/RT) in (cm^2-mole/cal-sec) of dyes and ratio of loss modulus to frequency (E"/ω) in (dynes/cm^2-cps) of film in tetrachloroethylene.

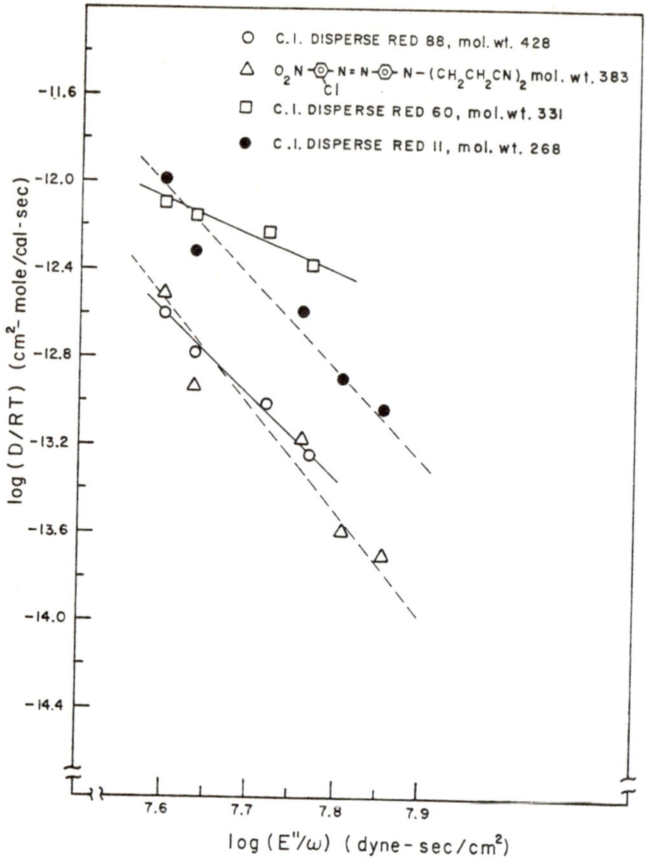

Fig. 7.24. Relationship between molecular mobility (D/RT) in (cm^2-mole/cal-sec) of dyes and ratio of loss modulus to frequency (E''/ω) in (dynes/cm^2-cps) of film in tetrachloroethylene.

volume, the values of D/RT over this temperature range decrease. Such a trend is compatible with explanations of disperse dye penetration based on the energy required to separate the surrounding polymer segments sufficiently for the dye molecule to pass [21]. The slopes of these lines vary from dye to dye, but generally indicate a value of B in the range from 1 to 4 for the equation ℓn (D/RT) = -B ℓn (E''/ω) + C'.

(2) Acrylic fibers. In a previous section on the relation between the dynamic properties of polyester fiber and dye diffusion behavior, it was found that theoretical relations derived by Fujita et al. [43] are applicable to the dyeing of polyester. The dynamic mechanical properties and dye diffusion behavior in acrylic fibers have

also been investigated [47]. In this study the dye used was a commercial grade of C.I. Basic Blue 22, Sevron Blue 2G (du Pont). The dyeing temperature was 100°C, and the liquor:fiber ratio was 40:1. A dye concentration of 5 g/l was used, and the pH of the dye bath was adjusted to 5.2 with acetic acid. The dyeing rates of four acrylic fiber samples were measured by techniques described in [46]. The fibers were as follows: No. 1, acrylonitrile—vinyl acetate copolymer containing 10.8% vinyl acetate; No. 2, acrylonitrile—vinyl acetate copolymer containing 7.5% vinyl acetate, stretched 300%; No. 3, acrylonitrile—vinyl acetate copolymer containing 9.2% vinyl acetate, stretched 300%; No. 4, same as No. 3, but stretched 400%.

The dynamic mechanical properties were measured at a constant frequency of 11 Hz with a Rheovibron viscoelastometer. All measurements were made with the sample in a 100% RH environment, since one was interested in the dynamic properties under dyeing conditions.

Figure 7.25 shows the dynamic modulus E', the loss modulus E", and the loss tangent at 0 and 100% RH as a function of temperature for sample 1. At 0% RH the temperature of the maximum of the tan δ peak for this sample occurs at 128°C, at a maximum tan δ value of 0.380. This α-dispersion is due to the amorphous regions in the structure and occurs in many semicrystalline polymers. At 100% RH the peak temperature of tan δ is shifted approximately 48°C lower than it is at 0% RH, but the maximum height of the tan δ peak is unchanged. A large shift is found for nylon 66 from dynamic data [12,46] and creep [65]. The shift for nylon is 15°C per percent of moisture. Since the water absorption of acrylic fibers at 90°C and 100% RH is about 3% [66], the shift is 16°C per percent of moisture. This suggests that the effect of moisture on the α-dispersion is due to a similar mechanism in these semicrystalline polymers.

The dye diffusion constant was calculated by the method used by Rosenbaum [67], which gives numerical results proportional to values obtained by the method used in work on nylon [46]. The equation is

$$D = \frac{C_t^2}{2\pi^{1/2}\ C_\infty^2\ A\omega^2 t}, \qquad (7.19)$$

where D = diffusion constant (cm²/sec), C_∞ = dye uptake at saturation (g of dye per cm² of fiber), C_t = dye uptake at time t (g of dye per g of fiber), A_ω = fiber surface area (cm² per g of fiber), and t = time (sec).

The dyeing-rate data are shown in Fig. 7.26. The dye uptake C_t is linear as a function of $t^{1/2}$, as predicted by the equation, and

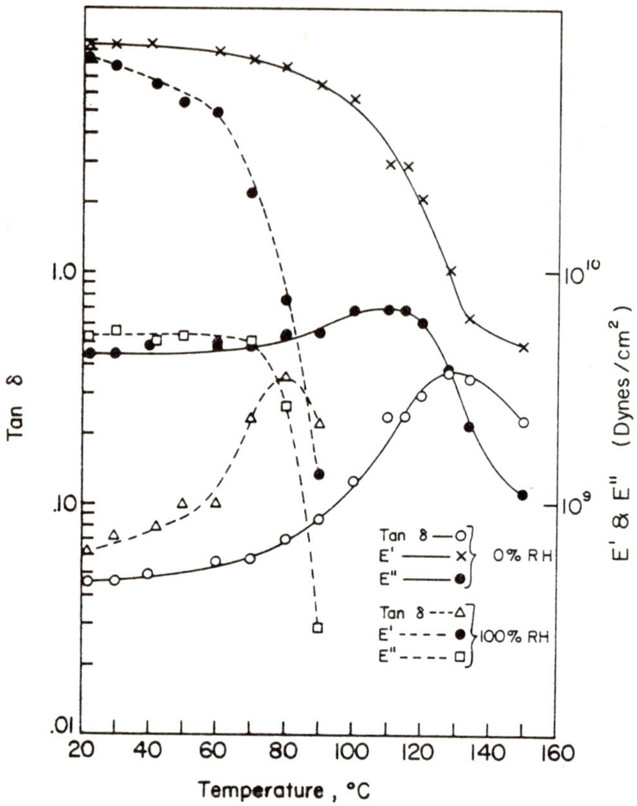

Fig. 7.25. Dynamic mechanical properties of acrylonitrile-vinyl acetate copolymer containing 10.8% vinyl acetate, as a function of temperature at 0 and 100% RH.

the line passes through the origin. The deviation from linearity in the top curve is caused by the approach to saturation. The relationship between E'' and D is shown in Fig. 7.27, plotted in the modified form of the equation of Fujita et al. [46].

As for nylon, the data appear to fit the equation. For the acrylic fiber $B_d = -2$, whereas for nylon $B_d = -4.5$. The value $B_d = 1$ was found by Fujita et al. in their studies of the diffusion of organic liquids in amorphous polymers. One possible cause is that E'' measurements are made parallel to the fiber axis, whereas dye diffusion takes place primarily perpendicular to it. The effect of orientation parallel to the fiber axis is probably different for E'' and D. The relation between E'' and D to acrylic fibers indicates

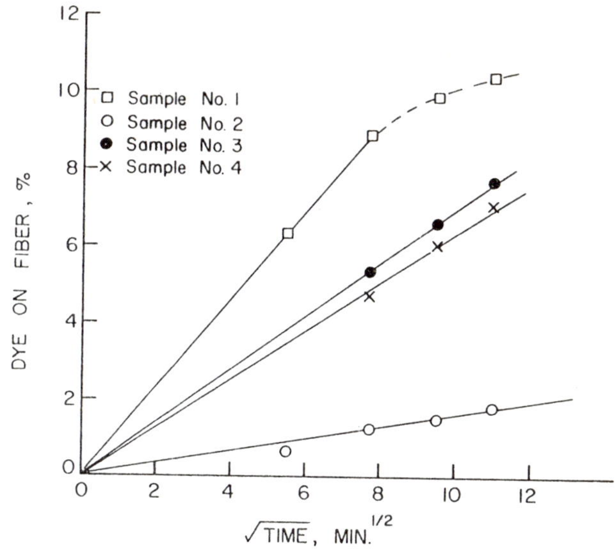

Fig. 7.26. Dye absorption as a function of time.

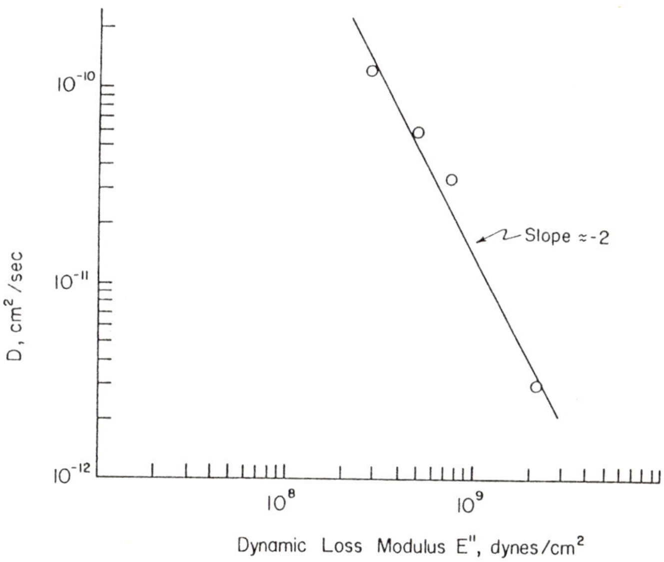

Fig. 7.27. Relation between diffusion constant and dynamic loss modulus.

that the dependence of dye diffusion on polymer-chain mobility is a general phenomenon, not limited to nylon and polyester.

Rosenbam [67] also reasoned that dye diffusion should be related to the mobility of the polymer chain segments, and he plotted the diffusion constant in acrylic polymers as a function of the tensile compliance (reciprocal of the initial modulus) and as a function of the fluidity (the reciprocal of the steady-state viscosity from creep measurements). The data appeared to follow a linear relationship in both cases, within the range of diffusion constants investigated.

6. BIOPOLYMERS

Dynamic mechanical analysis has been utilized to study the relation between structure and properties of biopolymers. Several investigators have reported the dynamic mechanical properties of collagen [68-71]. Dynamic study of collagen is necessary because the principal function of collagen is to provide support for tissues. The biological state of a tissue is often dependent upon the strength and integrity of its support, as is evident in the aging of collagen. Physical and mechanical characterization of collagen is important for another reason. Collagen and its derivatives have been developed for use as artificial membranes [72], tubes and sutures [73], and replacements for damaged corneas [74]. Quantitative comparison of the properties of synthetic or reconstituted materials with those of native collagen is essential for these applications. Collagen is also now being used in conjunction with nonthrombogenic surfaces and as enzyme supports.

Nguyen et al. [68] have examined the dynamic mechanical properties of the reconstituted beef tendon collagen which is prepared by an acidic solution of ficin-digested collagen. The details of this preparation are given in their publication [68]. Films were made from that solution by casting on a glass or poly(methyl methacrylate) surface. Films having different moisture contents were prepared by allowing them to equilibrate with an environment of fixed relative humidity or by drying at a fixed temperature for some time. A Rheovibron was used to measure dynamic mechanical properties over the range of temperature from -100°C to 230°C.

The sample chamber was continually swept with dry nitrogen to prevent condensation of moisture and to minimize thermal degradation at the higher temperatures. Typical sample dimensions were 0.075 X 4 X 50 mm. The dynamic moduli and the loss tangent (tan δ) were obtained at five different moisture contents (MC = 2, 10, 12, 18, and 23%) on this collagen. The dynamic modulus of 2% MC sample showed the least temperature dependence in the temperature range from -100 to 210°C. The variation of the dynamic modulus in this temperature range was from 6.0 X 10^9 to 3.0 X 10^9 dyn/cm^2. However, the modulus and its dependence upon temperature are clearly correlated with moisture content. This is shown by the modulus changes in temperature range from -60°C to 60°C. The moduli of the 18% and 23% MC samples decrease with increasing temperature in the temperature range from -100°C to 18°C, and then increase beyond 18°C up to about 60°C. The modulus is nearly constant above 60°C until about 218°C, where it decreases very sharply. The loss-tangent (tan δ) data show three loss peaks in the temperature range from -100 to 210°C. The peak temperatures are about 50°C (α), 10°C (β), and -70°C (γ).

The magnitudes of three loss peaks are: α-peak, tan δ_{max} = 0.07; β, 0.08; γ, 0.06. The β-peak indicates high energy loss in cyclic deformation of collagen. The frequency dependence of the loss tangent as a function of temperature has been investigated by Nguyen et al. [68] on the same collagen (18% MC). They found that the magnitudes of the loss peaks α and β decrease with decreasing moisture content, and suggest that this behavior is probably associated with "melting" and desorption of water. Similar conclusions have been drawn by Dehl [75] after looking at the effect of bound water in wet collagen by NMR relaxation studies. The effect of water on the loss tangents of human diaphragm tendon collagen has also been noted by Baer et al. [71]. Hiltner et al. have studied [70] the effect of water on the dynamic modulus and loss factors of various hydrophilic synthetic polypeptides.

The sharp reduction in dynamic modulus at 218°C of the 2% MC sample can be related to the melting of collagen. Okamoto and Saeki [76] and Oth suggested that the T_m of dry collagen is about 200°C on the basis of direct observation of shrinkage of uncrosslinked collagen on a hot stage in a polarizing microscope. Puett [76,77] suggested that the endothermic behavior of collagen and other polypeptides observed in differential thermal analyses (DTA) could be related to their melting or denaturation.

Chien et al. also have studied the dynamic mechanical properties of rat-tail tendons (RTT), enzyme-solubilized collagen membranes (ESC), AKM-23 dialysis membranes, and gelatin film. Measurements were made at 3.5, 11, and 110 Hz from -160 to 220°C with a Rheovibron.

The dynamic modulus for these samples decreases with increasing temperature, reaching a minimum between 0°C and 30°C. This transition is attributed to the devitrification of water bound to the protein

[78-80]. The transition is sharp for rat-tail tendons and very broad for AKM-23 [this membrane contains 63% molecularly dispersed collagen fibril, 27% enzyme-solubilized collagen membranes (ESC), and 10% glycerine as plasticizer]. In both these materials it is more pronounced than in the others. The dynamic modulus of AKM-23 in the temperature range from -150°C to -130°C is about 5.0 X 10^{10} dyn/cm^2; then it decreases sharply from 5.0 X 10^{10} to 2.0 X 10^9 dyn/cm^2 in the temperature range from -100°C to 0°C. However, it increases with increasing temperature from 0°C to 150°C. At 160°C it decreases sharply again. In the case of rat-tail tendons (RTT), the dynamic modulus is about 9.0 X 10^{10} dyn/cm^2 around -100°C, but becomes about 1.0 X 10^{10} dyn/cm^2 at 10°C. It increases with increasing temperature beyond 10°C. The modulus at 80°C is about 7.0 X 10^{10} dyn/cm^2.

The modulus recovery is greater than 80% for RTT. This tensile strength originates primarily in the supermolecular structure of oriented fibrils and is less sensitive to the loss of moisture. These samples show three loss transitions. The lowest-temperature (γ) loss peak occurs between -70 and -100°C. The magnitude of this peak is about 0.21 for AKM 23 and about 0.03 for RTT. Chien concludes this peak is due to the γ-mechanism seen in polyamides. Koleski and Faucher [80] reported that addition of water to gelatin decreased the width and increased the magnitude of the peak without shifting its position.

The loss-tangent curve of RTT shows a loss peak at 0°C. Its intensity is higher than that of the other loss peaks. This transition is related to the devitrification temperature. At 210-220°C for most samples and at 170°C for AKM-23, the materials begin to show signs of flow. Some investigators have ascribed these changes to a melting process [76,81-84].

Hiltner et al. [70] have studied a series of copolymers of L-leucine and L-glutamic acid as a model system for collagen. The dynamic mechanical behavior was investigated as a function of water content. Their results suggest a spectrum of water-polymer interaction, with the two predominant types represented by "tightly bound" and "loosely bound" water. Each gives rise to a characteristic relaxation peak.

Dynamic mechanical properties of keratin and polypeptides have been studied by several investigators [85-88]. These studies conclude that dynamic mechanical techniques can provide meaningful correlation between molecular structure and mechanical properties.

REFERENCES

1 T. Murayama and B. Silverman, J. Appl. Polym. Sci., 19 (1975) 1695.
2 J. P. Bell and T. Murayama, J. Appl. Polym. Sci., 12 (1968) 1795.
3 J. H. Dumbleton, J. P. Bell, and T. Murayama, J. Appl. Polym. Sci., 12 (1968) 2491.
4 J. P. Walters, Textile Res. J., 41 (1971) 857.
5 T. Murayama and A. A. Armstrong, Jr., J. Appl. Polym. Sci., 12 (1974) 1211.
6 M. Takayanagi, Mem. Fac. Eng., Kyushu Univ., Japan, 23 (1963) 1.
7 D. L. MacLean and T. Murayama, Trans. Soc. Rheol., 18:2 (1974) 237.
8 Y. Murai and N. Kuroda, Seni-Gakkaishi, 29 (1973) T-38.
9 H. G. Zachmann and W. Scherman, Kolloid-Z. Z.-Polym., 241 (1970) 916.
10 A. B. Desai and G. L. Wilkes, Textile Res. J., 45 (1975) 173.
11 K. Suzuki, T. Asano, I. Ikeda, and I. Kido, Seni-Gakkaishi, 30 (1974) T-413.
12 A. E. Woodward, J. M. Crissman, and J. A. Sauer, J. Polym. Sci., 44 (1960) 23.
13 J. H. Dumbleton and T. Murayama, Kolloid-Z. Z.-Polym., 238 (1970) 410.
14 K. V. Datyes, S. C. Pitkar, and U. M. Purso, Textilveredlung, 6 (1971) 593.
15 F. O. Harris and T. H. Guion, in Textile Solvent Technology--Update 73, Am. Assoc. of Textile Chem. and Color., Research Triangle Park, NC, 1973, p. 163.
16 W. S. Perkins and D. H. Hall, in Textile Solvent Technology--Update 73, Am. Assoc. of Textile Chem. and Color., Research Traingle Park, NC, 1973, p. 49.
17 Z. Morita, R. Kobayashi, K. Uchimura, and H. Motomura, J. Appl. Polym. Sci, 19 (1975) 1095.
18 A. Katayama, Y. Kitano, and N. Kuroki, Seni-Gakkaishi, 28 (1972) 302.
19 T. Haga and H. Ishibashi, Seni-Gakkaishi, 29 (1973) T489.
20 K. Bredereck and E. Koch, Melliand Textilber. (Engl. edn.), 3 (1974) 118.
21 A. S. Ribnick and H. D. Weigmann, Textile Res. J., 43 (1973) 316.
22 B. H. Knox, H. D. Weigmann, and M. G. Scott, Textile Res. J., 45 (1975) 203.
23 E. L. Lawton and T. Murayama, J. Appl. Polym. Sci., 20 (1976) 3033.
24 A. S. Ribnick, H. D. Weigmann, and L. Rebenfeld, Textile Res. J., 43 (1973) 176.
25 J. H. Dumbleton and T. Murayama, Kolloid-Z. Z.-Polym., 220 (1967) 41.
26 M. A. Harrison, P. H. Morgan, and G. S. Park, Eur. Polym. J., 8 (1972) 1361.
27 H. Jedraszczyk, Dent. Textiltech, 21 (1971) 233.
28 K. Bredereck, H. Dolmetsch, E. Koch, and R. Schoner, Melliand Textilber. (Engl. edn.), 4 (1975) 31.
29 B. Illers and H. Breuer, J. Colloid Sci., 18 (1963) 1.
30 G. W. Becker and H. Oberst, Kolloid-Z. Z.-Polym., 152 (1975) 1.
31 G. King, J. Colloid Sci., 2 (1947) 551.
32 K. H. Illers, Makromol. Chem., 38 (1960) 168.
33 A. E. Woodward, J. A. Sauer, C. W. Deely, and D. E. Kline, J. Colloid Sci., 12 (1957) 363.
34 J. B. Jackson, Polymer, 10 (1965) 159.
35 M. Takayanagi, Viscoelastic Properties of Crystalline Polymers, Mem. Fac. Eng., Kyushu Univ., Japan, 26 (1963) 1.

36 H. Jacobs and E. Jenckel, Makromol. Chem., 43 (1961) 132.
37 K. H. Illers, Makromol. Chem., 38 (1960) 168.
38 J. B. Jackson, Polymer, 10 (1969) 159.
39 J. Janacek and J. Kolarik, J. Polym. Sci., C16, 279 (1967) 441.
40 J. Janacek, J. Polym. Sci., C23 (1968) 373.
41 M. Baccaredda, E. Butta, V. Frosini, and S. de Petris, J. Polym. Sci., A25 (1967) 1296.
42 Y. O. Zari, R. H. Chow, and J. K. Gillham, Polym. Prepr., 18-1 (1977) 649.
43 H. Fujita, A. Kishimoto, and K. Matsumoto, Trans. Faraday Soc., 56 (1960) 424.
44 F. Bueche, Physical Properties of Polymers, Interscience, New York, 1962.
45 J. H. Dumbleton, J. P. Bell, and T. Murayama, J. Appl. Polym. Sci., 12 (1968) 2491.
46 J. P. Bell, J. Appl. Polym. Sci., 12 (1968) 627.
47 J. P. Bell and T. Murayama, J. Appl. Polym. Sci., 12 (1968) 1795.
48 T. Masuko, M. Karasawa, K. Yonetake, and S. Okajima, Seni-Gakkaishi, 32 (1976) T-496.
49 H. Kojima and T. Iijima, J. Soc. Dyers Colour., 91 (1975) 103.
50 M. L. Williams, R. F. Landel, and J. D. Ferry, J. Am. Chem. Soc., 77 (1955) 3701.
51 A. K. Doolittle, J. Appl. Phys., 22 (1951) 1471.
52 H. Fujita, Fortschr. Hochpolym.-Forsch., 3 (1961) 1.
53 M. H. Cohen and D. Turnbull, J. Chem. Phys., 31 (1959) 1164.
54 J. Rosenbaum, J. Polym. Sci., A3 (1965) 1949.
55 Z. Gur-Arieh and W. Ingramells, J. Soc. Dyers Colour., 90 (1974) 8.
56 W. Ingamells, R. H. Peters, and S. R. Thornton, J. Appl. Polym. Sci., 17 (1973) 3733.
57 R. H. Peters and W. Ingamells, J. Soc. Dyers Colour., 89 (1973) 397.
58 K. Kitamura, Seni-Gakkaishi, 30 (1974) T-471.
59 J. H. Dumbleton and B. B. Bowles, J. Polym. Sci., A, 4 (1966) 951.
60 H. P. Klug and L. E. Alexander, X-Ray Diffraction Procedures, Wiley, New York, 1954.
61 D. Ya. Tsvankin, Vysokomolekul. Soedin., 6 (1964) 2078, 2083.
62 J. H. Dumbleton, J. Polym. Sci., A-2, 6 (1968) 795.
63 A. V. Hill, Proc. R. Soc., 104B (1928) 39.
64 P. R. Pinnock and I. M. Ward, Trans. Faraday Soc., 62 (1966) 1308.
65 W. H. Howard and M. L. Williams, Textile Res. J., 36 (1966) 691.
66 R. Jeffries, J. Textile Inst., 51 (1960) T411.
67 S. Rosenbam, J. Appl. Polymer Sci., 7 (1963) 1225.
68 A. Nguyen, Ban The Vu, and G. L. Wilkes, Biopolymers, 13 (1974) 1023.
69 J. C. W. Chien and E. P. Chang, Biopolymers, 11 (1972) 2015.
70 A. Hiltner, J. M. Anderson, and E. Baer, J. Macromol. Sci.--Phys., B8(3-4) (1973) 431.
71 E. Baer, R. Kohn, and Y. S. Paysir, J. Macromol. Sci.--Phys., B6 (1972) 761.
72 A. L. Rubin and K. H. Stenzel, Biomaterial (1969) 157.
73 A. L. Rubin, T. Miyata, and K. H. Stenzel, J. Macromol. Sci.--Chem., A3 (1969) 113.
74 M. W. Dunn, T. Nishihara, K. H. Stenzel, A. W. Branwood, and A. L. Rubin, Science, 157 (1967) 1329.
75 R. E. Dehl, Science, 170 (1970) 736.
76 D. Puett, Biopolymers, 5 (1967) 327.
77 A. Ciferri, L. V. Rajagh, and D. Puett, Biopolymers, 3 (1965) 461.
78 A. A. Khanagor and S. P. Gakuda, Biofizika, 14 (1969) 796.

79 G. N. Ramachandran and R. Chandrasekharan, Biopolymer, 6 (1968) 1649.
80 J. V. Koleske and J. A. Faucher, J. Phys. Chem., 69 (1965) 4240.
81 N. D. Felix, M. A. McDowall, and H. Eyring, Textile Res. J., 33 (1963) 465.
82 Y. Okamoto and K. Saeki, Kolloid-Z. Z.-Polym., 194 (1964) 124.
83 I. V. Yannas and A. V. Tobolsky, Eur. Polym. J., 4 (1968) 257.
84 J. E. Jolley, Photogr. Sci. Eng., 14 (1970) 169.
85 E. Menefee and G. Yee, Textile Res. J., 35 (1965) 801.
86 P. Mason, Nature, 212 (1966) 719.
87 M. Fenghelman, Applied Polymer Symposium No. 18, 1971, p. 757.
88 M. Druhala and M. Fenghelman, Colloid-Z. Polym. Sci., 252 (1974) 381.

Conversion factors for moduli, stress, and viscosity

To convert from	To	Multiply by
Newtons/meter^2 (N/m^2)	dynes/cm^2	10.00
dynes/cm^2	Newtons/meter^2 (N/m^2)	0.100
psi	N/m^2	6.895×10^{3}
N/m^2	psi	1.450×10^{-4}
dynes/cm^2	kg/cm^2	1.020×10^{-6}
dynes/cm^2	kg/sq. mm	1.020×10^{-8}
kg/sq. mm	dynes/cm^2	9.806×10^{7}
dynes/cm^2	psi	1.450×10^{-5}
psi	dynes/cm^2	6.895×10^{4}
psi	kg/sq. mm	7.03×10^{-4}
kg/sq. mm	psi	1.422×10^{3}
dynes/cm^2	atmospheres	9.869×10^{-7}
atmospheres	dynes/cm^2	1.013×10^{6}
atmospheres	N/m^2	1.013×10^{5}
psi	atmospheres	6.81×10^{-2}
dynes/cm^2	bars	1.00×10^{-6}
psi	bars	6.895×10^{-2}
g/denier	dynes/cm^2	$8.83 \times 10^{8}\ \rho$*
g/denier	psi	$1.28 \times 10^{4}\ \rho$*
bars	N/m^2	1.00×10^{5}
kg/cm^2	N/m^2	9.807×10^{4}
poise	N·S/m^2	1.000×10^{-1}
stokes	m^2/S	1.000×10^{-4}
dynes	newtons (N)	1.000×10^{-5}

*ρ = density

AUTHOR INDEX

Aggarwalt, S. L., 58
Ajroldi, G., 94
Aklonis, T. T., 12, 158
Alcock, T. C., 35
Alexander, L. E., 95, 158, 222
Alfrey, T., 34
Alptekin, A. T., 137
Anderson, J. M., 222
Andrews, R. D., 148
Antkowiak, T. A., 58
Armstrong, A. A., 221
Asano, T., 221
Ashton, J. E., 175
Aso, S., 136, 138

Baccaredda, M., 222
Baer, E., 94, 219
Bagley, E. B., 58
Ballow, J. W., 57, 137, 176
Bay-Sung, Hsu, 67, 94
Beaman, R. G., 34
Becker, G. W., 221
Behrens, E., 138
Bell, J. P., 66, 89, 94, 177, 221
Berge, J. W., 57
Billmeyer, F. W., 35
Blaine, R. L., 59
Blake, M. P., 34, 138
Boerma, J., 95
Bowles, B. B., 95, 148, 222
Boyer, R. F., 34, 46, 60, 64, 94
Brailsford, A. D., 158
Branwood, A. W., 222
Brauer, J., 34
Bredereck, K., 221
Brederode, R. A. Van, 57
Bree, H. W., 138
Breuer, H., 201
Broutman, L. J., 138
Brown, J. E., Jr., 96
Brownlee, E. L., 112, 137
Broyer, E., 58
Bruener, H., 68, 77
Bryant, C. W., 35
Buchanan, D. R., 158
Buchdahl, R., 57, 96, 138
Bueche, F., 222
Bunn, C. U., 35
Butta, E., 222

Cerro, R. L., 58
Chaikin, M., 137
Chamberlain, N. H., 137
Chandrasekharan, R., 223
Chang, E. P., 222
Chen, A. C. F., 135, 138
Cheng, C. C., 137
Chien, J. C. W., 219
Chow, R. H., 222
Christensen, R. M., 34, 139
Church, W. H., 176
Ciferri, A., 222
Clark, E. S., 34
Clark, S. K., 34
Cocks, G. G., 57
Cohen, M. H., 204
Collins, E. A., 58
Corradini, P., 35
Cox, W. P., 79, 156
Craver, J. K., 59
Crissman, J. M., 95, 177, 221
Crystal, R. G., 58, 177

Darin, S. R., 158
Datyes, K. V., 221
Davenport, N. E., 95
Davis, S. S., 58
Davis, W. M., 59
Deeley, C. W., 177, 221
Dehl, R. E., 219
Den Hartog, J. P., 59
Desai, A. B., 221
Deutsch, K., 62, 94
Dillon, J. H., 57
Di Marzio, E. A., 95
Doak, K. W., 94
Dodge, C. W. H., 95
Dolmetsch, H., 221
Doolittle, A. K., 34, 204
Doolittle, D. B., 34
Druhala, M., 223
Drumm, M. F., 95
Duckett, K. E., 137
Duffin, D. J., 138
Dumbleton, J. H., 34, 77, 85, 137, 148, 161, 221
Dunn, M. W., 222

Ender, D. H., 148
Erhardt, P. F., 41, 177
Eyring, H., 223

Farrow, G., 67, 94
Faucher, J. A., 220
Felix, N. D., 223
Fenshelman, M., 223
Feodosev, V. I., 121
Ferry, J. D., 18, 57, 60, 139, 143
Fielding-Russell, G. S., 57
Fischer, E. W., 35
Fitzgerald, E. R., 57
Flory, P. J., 34, 95
Follmert, B., 35
Fox, T. G., 95
Frankenfeld, F., 138
Frosini, V., 222

Fujino, K., 57
Fujita, H., 203
Fukada, E., 42

Gakuda, S. P., 222
Garbuglio, C., 94
Gashtol'd, N. O. S., 64
Gehman, S. D., 57
Geil, P. H., 22, 94
Gent, A. N., 57
Gillham, J. F., 45, 177, 222
Glick, R. E., 177
Goltzev, D. I., 137
Goodman, L. E., 121
Goswami, B. C., 138
Gray, R. W., 95
Grote, W. A., 58
Gruver, J. T., 138
Guion, T. H., 221
Gupta, R. P., 175
Gupta, V. B., 177
Gur-Arieh, Z., 204

Hadley, D. W., 176
Haga, T., 221
Hail, D. H., 221
Halboth, H., 95
Halpin, J. C., 34, 137, 138
Hamburger, W. J., 137, 176
Hamme, R. N., 34, 138
Hara, T., 176
Harris, F. O., 221
Harris, J., 58
Harrison, M. A., 221
Hartman, B., 177
Hay, I. L., 176
Hazony, Y., 58
Hearles, J. S. W., 100, 137
Hearmon, R. F., 175
Heijboer, J., 62, 94
Heinze, H. D., 95
Hermann, K., 35
Hermans, J. J., 138
Heydemann, P., 58
Hideshima, T., 158
Hill, A. V., 206
Hillier, K. W., 176
Hiltner, A., 219
Hoff, E. A. N., 62
Holland, V. F., 31
Holliday, L., 137
Horio, M., 57, 58
Hosemann, R., 29
Howard, W. H., 141, 222
Hubbard, I. W., 95
Hunt, B. I., 62, 64

Iijima, T., 204
Ikeda, I., 221
Illers, B., 201
Illers, K. H., 57, 68, 77, 222
Imada, K., 176
Ingramells, W., 222
Isaksen, R. A., 94
Ishibashi, H., 221
Iwayanagi, S., 147
Izard, E. F., 95

Jackson, J. B., 221
Jacobs, H., 222
Janacek, J., 222
Jedraszczyk, H., 221
Jeffries, R., 222
Jenckel, G., 57, 95, 222
Johnson, J. F., 35
Jolley, S. E., 223

Kelbe, D. H., 116, 137
Kajiyama, T., 42, 176
Kanig, G., 94
Kainradl, P., 138
Karasawa, M., 222
Katayama, A., 221
Kaufmann, G., 138
Kawaguchi, T., 94
Kawai, H., 57
Ke, B., 34
Keil, M., 35
Keith, H. D., 35
Keller, A., 31, 94, 176
Kemp, R. A., 58
Kenyon, A. S., 42
Kerwin, E. M., Jr., 138
Khanagor, A. A., 222
Kido, I., 222
Kilian, H. G., 95
King, G., 221
Kinoshita, R., 136
Kishimoto, A., 203
Kitamura, K., 204
Kitano, Y., 221
Kline, D. E., 67, 94, 177
Klug, H. P., 95, 158, 222
Knox, B. H., 193
Kobayashi, R., 221
Koch, E., 221
Kodama, K., 95, 138
Koga, K., 42
Kohn, R., 222
Kojima, H., 223
Kolarik, J., 222
Kolb, H. J., 95
Koleske, J. V., 220
Kolsky, H., 34, 176
Koppelmann, J., 57
Kosten, C. W., 138
Kraus, G., 138
Krigbaum, W. R., 76
Krock, R. H., 138
Kuroda, N., 199
Kuroki, N., 221
Kyker, G. S., 58

Lamb, G. E. R., 138
Landel, R. F., 18, 138, 222
Lawton, E. L., 137, 196
Leaderman, H., 17, 158
Lee, G. F., 177
Levina, Z. M., 137
Lewis, T. B., 138
Lindenmeyer, P. H., 34
Livingston, D. I., 57, 96
Lord, P., 57
Loshaek, S., 95
Lustig, S., 35

MacKnight, W. J., 34, 158
MacLean, D. L., 221
Macosko, C. W., 58
Mandelkern, L., 34, 95
Marei, A. I., 64
Mark, H., 35
Martin, G. M., 95
Marvin, R. S., 57
Mason, P., 223
Massa, D. J., 41, 177
Masuko, T., 222
Mater, J., 177
Matsui, J., 137
Matsumoto, K., 203
Matsuo, T., 94
McCrum, N. G., 34, 60, 95, 150
McDowall, M. A., 223
McMaster, R. C., 59
McIntosh, J., 67, 94
Menefee, E., 222
Meredith, R., 67, 94, 176
Merz, E. H., 94
Miller, R. L., 68
Milloway, W. T., 58
Minami, S., 96
Mitchell, W. S., 34, 138
Miyagi, A., 95
Miyamoto, T., 95, 138
Miyata, T., 222
Mohr, J. G., 138
Montgomery, D. J., 58
Morgan, H. M., 176
Morgan, P., 138
Morgan, P. H., 221
Morgan, R. J., 57
Morita, Z., 221
Morris, E. L., 158
Morse, P. J., 58
Moseley, N. W., 176
Motomura, H., 221
Murai, Y., 199
Murayama, T., 34, 41, 81, 85, 98, 111, 158, 161, 178
Mussatti, F. G., 58

Nakayama, C., 57
Nakajima, N., 58
Nakene, H., 158
Naghazadeh, J., 64
Natta, G., 27
Nederveen, C. T., 57, 138
Newman, S., 79, 155
Nguyen, A., 218, 219
Nielsen, L. E., 34, 57, 60, 64, 66, 131, 156, 175
Nishihara, T., 222
Nye, J. F., 175

Oberst, H., 138, 221
Okajima, S., 222
Okamoto, Y., 219
Okano, K., 158
Oleesky, S. S., 138
O'Malley, J. J., 58, 177
Onogi, S., 57

Padden, F. J., Jr., 35
Pagano, N. J., 34, 137
Pain, H. S., 34
Panovko, G. Ya, 121
Park, G. S., 221
Passagha, E., 95
Paul, D. R., 58
Paysir, Y. S., 222
Pennings, A. J., 35
Perkins, W. S., 221
Perry, E., 138
Peters, R. H., 222
de Petris, S., 222
Pezzin, G., 94
Pinnock, P. R., 67, 94, 159
Pitkar, S. C., 221
Platt, M. M., 137
Plazek, D. J., 57
Porod, G., 94
Porter, R. S., 34
Poules, J. G., 62, 94
Price, F. P., 68, 94
Puett, D., 219
Purso, U. M., 221
Putti, M. R., 95

Radcliffe, S. V., 94
Raff, R. A. V., 94
Rajagh, L. V., 222
Ramachandra, G. N., 222
Raumann, G., 159, 175
Rayford, M. C., 58
Rayleigh, J. W., 58
Read, B. E., 34, 60
Rebenfeld, L., 221
Reddish, W., 62, 94, 155
Reshetov, D. N., 121, 137
Rhodes, M. B., 68
Ribnick, A. S., 197
Richardson, E. G., 138
Richmond, P. G., 138
Riding, G., 138
Riley, M. B., 137

Rodrigues, F., 57
Roe, R. J., 76
Roller, M. B., 57
Rosenbaum, J., 204, 218, 222
Rubin, A. L., 222
Ruzicka, J. E., 137

Saeki, K., 219
Saito, N., 158
Samuels, R. J., 35
Samuels, R. T., 95, 149
Sandiford, D. J. H., 62
Sato, K., 158
Sauer, J. A., 34, 67, 95, 177, 221
Saunders, D. W., 177
Scherman, N. W., 201
Schnell, G., 95
Schoner, R., 221
Schreiner, S. A., 138
Schultz, A. B., 138
Schwarzl, F. R., 131, 137
Schwippert, G. A., 138
Scott, M. G., 221
Scriven, L. E., 58
Seto, T., 176
Shatzik, T. F., 64
Shen, M., 34, 158
Shibayama, K., 94, 138
Shroff, R. N., 57
Sidorouitch, E. A., 64
Silverman, B., 81, 95, 178
Silverman, J. S., 137
Sinnott, K. M., 62, 158
Smith, J. C., 57, 76, 95, 176
Smith, T. L., 138
Southern, J. H., 58
Stachurski, Z. H., 161, 177
Stadnicki, S. J., 58
Standinger, H., 23
Starita, J. M., 58
Stein, R. S., 68, 149
Stenzel, K. H., 222
Stewart, L. E., 81
Strakhov, G. N., 137
Strella, S., 58
Struik, L. C. E., 137
Stuart, Ann, 35
Stuart, H. A., 35
Suzuki, K., 221

Takayanagi, M., 35, 37, 42, 60, 63, 90, 137, 153, 160, 221
Takeyama, T., 137
Taylor, D. L., 59
Taylor, G. R., 158
Taylor, N. W., 58
Thomas, A. M., 67, 164
Thompson, A. B., 155
Thornton, S. R., 222
Till, R. H., 35
Timoshenko, S., 34, 177
Tobolsky, A. V., 18, 95, 96, 158, 223
Tokita, N., 57
Treloar, L. R. G., 100
Tsai, S. W., 34, 137, 138
Tsvankim, D., 80, 222
Turnbull, D., 204

Uberreiter, K., 64, 95
Uchimura, K., 221
Uematsu, I., 76
Uematsu, Y., 76
Uemura, S., 96
Utracki, L., 64

Vadimsky, R. G., 35
Van der Mark, J. M. A. A., 35
Van der Wal, C. W., 57, 137
Van Hoorn, H., 95
Van Schooten, J., 95
Vaughan, M. F., 35
Vokulonskaya, I. I., 138
Volkova, T. A., 138
Vrancken, M. N., 57

Wall, R. A., 138
Wallace, D. A., 58
Walters, J. P., 221
Ward, D. W., 58
Ward, I. M., 57, 67, 153, 159, 160, 176, 222
Ward, I. W., 34
Weighmann, H. D., 138, 197
Weissert, F. C., 58
Wetton, R. E., 57
White, J. L., 137
Whitehurst, E. A., 59
Whitley, G. S., 35
Whitney, J. M., 138, 175
Wiley, R. H., 34
Wilkes, G. L., 221
Wilkinson, W. L., 58
Willbourn, A. H., 64
Willett, P. R., 138
Williams, G., 34, 60
Williams, H. L., 135
Williams, M. L., 18, 137, 141, 222
Wolf, K. A., 95, 96
Wolfe, J. M., 176
Woo, L., 58
Woods, D. W., 67, 155
Woodward, A. E., 34, 86, 95, 177, 189, 201, 222
Wunderlich, B., 22, 95

Yamamoto, K., 57
Yannas, I. V., 223
Yee, G., 223
Yoshino, M., 37, 94
Yonetake, K., 222

Zari, Y. O., 222
Zachmann, H. G., 201
Zorowski, C. F., 59, 98, 111, 137, 158
Zosel, A., 58
Zubov, P. I., 138
Zuikker, C., 138

SUBJECT INDEX

ABS resin, 90
Acetate yarn, 102
Acoustics, 134, 135
Acrylic fibers, 175, 214
Adhesion, 111, 116, 125
Alpha peak, 63, 64, 85
Amorphous volume, 76
Amorphous state, 28
Anisotropy, 159, 161
Annealing, 67, 71
Asphalt-based deadeners, 134
Assembly factor, 106

Bending, 171
Beta peak, 64
Biopolymers, 218
Birefringence, 149
B&K audio-frequency
spectrometer, 135
complex modulus
apparatus, 53
Boltzmann superposition, 9, 140
Bonding characterization, 120
Bond strength, 108, 109

Chain folds, 31
Collagen, 218, 219
Composite, 124, 130
Complex representation, 5
compliance, 5
modulus, 8, 124
piezoelectric stress
constant, 42
equivalent model, 91
Compression, 164
Conformation, 28
Copolymers, 25, 90
Coulomb-friction model, 122
Crankshaft mechanism, 64
Crosslinking, 25, 86
Crystallinity, 67
Crystal size, 68, 71

Damping peak, 73
Deadeners, 134
Diffusion, 190, 202, 204
Dispersion, 61
Doolittle equation, 18, 19
Dyeing, 189, 203
absorption, 206
rate measurement, 205
Dynamic adhesion test, 111
Dynamic mechanical analyzer, 51
Dynamic mechanical
properties, 23
bending, 171
boiling water, effect of, 206
copolymers, 90
composite materials, 116
compression, 164
crosslinking, effect of, 87
crystallinity, effect of, 68
gas, effect of, 180
liquid, effect of, 181
molecular weight effects, 81
moisture, effect of, 200
nylon, 65, 161, 142
polyblends, 89
polyester, 23, 167, 170, 174
strain, effect of, 156
shear, 167
solvent, effect of, 189
temperature and frequency
effects, 21
viscous drag effect, 187

Epoxy, 86
Equivalent, viscous damping, 103
Equation of twisted model, 104

Fiber-to-fiber Coulomb parameter,
106, 107
Fiber-filled composite, 131, 132
Fiber orientation in composite, 132
Filler, 131
Flory equation, 17
Free radical, 24
Fringed-micelle model, 29
Friction, 108

Gamma peak, 64
Gas cell, 178, 179
Glass transition, 11, 64
Graft bonds, 90

Heat generation, 10
Heterogeneity, 92
High frequency wave propagation, 102
H-7 dip, 113
Humidity effect, 201

Interface, 98
Interface energy dissipation, 125
Interfilament friction, 108
In liquid, 186, 189, 191, 192
Internal friction, 2, 5

Linear viscoelastic behavior, 6
Liquid medium, 183
Logrithmic decrement, 45
Logrithmic rule of mixture, 67
Long period, 69
Loss factor, 63

Matrix-filament interface, 121
Maxwell element, 5, 9
Mechanical equivalent model, 90
Mechanical spectrometer, 48

Melting point, 12
Micro-Brownian motion, 11, 61
Moisture contents, 202
Molecular orientation, 67, 78
Molecular weight, 23, 25, 81, 89

Navier-Stokes equation, 184
Noise, 2, 134
Nonwoven fabrics, 106
Nylon, 6, 65
Nylon 66, 65, 81, 142, 161, 164, 174, 189

Orientation parameter, 79, 80
Organic liquids, 189

Parallel and perpendicular moduli, 160, 163
Phenomenological theory, 140
Polybutadiene rubber, 166
Polyethylene, 23, 24, 31
Polyethylene terephthalate, 23, 167, 170, 174, 191, 192, 204
Polymer-solvent system, 17
Polymerization, 23
Polymer blends, 90
Polyurethane elastomer, 166, 170
Polyvinyl alcohol (PVA), 201
Predicting effective dynamic modulus, 186
Primary dispersion, 61
Pseudoviscous effect, 107

Rat-tail tendson, 220
Relaxation modulus, 141, 143
Relaxation spectra, 65
Relaxation times, 8, 21
Resonance peak, 47
RFL dip, 113
Rheovibron, 36
- DDV-III-B, 42
- automated system, 42
- piezoelectric system, 42
- modified techniques, 166, 169, 172

Secondary dispersion, 61, 62
Segmental motion, 201
Shear complex modulus, 5
Shear modulus of crosslinked rubber, 88
Sherrer formula, 68
Shift factor, 18, 144, 152
Shift in E''_{max}, 75, 77
Shish kebab, 31
Shrinkage, 199
Sinusoidal excitation, 4
Small angle scattering, 69
Solid-state polycondensation (SSP), 81
Solvent dye, 213
Sonic technique, 54, 99
Sound generation, 136
Spherulites, 28
ST-AN copolymer, 90
Stereoregularity, 28
Stress-relaxation modulus, 18
Stress relaxation, 141
- apparatus, 150
- of nylon 66, 142
- of PET, 151

Stress-strain, 156
Structural alteration, 199
Structural damping, 134

$T_{\ell\ell}$ relaxation, 64
Tan δ peak, 73
T_g shift, 87
Three-lumped mechanical elements, 106, 107
Time-temperature superposition, 17, 140
Tire cord-rubber system, 111
Torsion pendulum, 43, 167
Torsional braid analysis, 45, 167
Twisted yarn, 102, 105

Vibration, 3
Vibrating reed, 47
Viscoelasticity, 139
Voigt element, 5
Volume fraction, 126, 127

Wave propagation, 3
Weissenberger Rheogoniometer, 48
WLF equation, 18

X-ray diffraction, 68

Yarn strain, 101